Cellulose - Biobased Solutions for Society

Edited by Alejandro Rodríguez

Published in London, United Kingdom

Cellulose - Biobased Solutions for Society
http://dx.doi.org/10.5772/intechopen.1005959
Edited by Alejandro Rodríguez

Contributors

Alejandro Rodríguez Pascual, Ana Balea, Angeles Blanco, Anurag Kashyap, Behrooz Maleki, Bhargavi Pathak, Carlos Negro, Elena Fuente, Hongyu Xu, Jose Luis Sanchez-Salvador, K. Sivaprakash, M. Concepcion Monte, M. Lavanya, Madhu Sasikala, Mariama Alidu, Munmi Sarma, Pouya Ghamari Kargar, Remedios Montenegro, Runjun Sarma, Samaneh Sedigh Ashrafi, Shahida Choudhury, Sukanya Saikia, Symone Alexander, V. Ramesh Prabu, Zoilo González

First published in London, United Kingdom, 2025 by IntechOpen
IntechOpen is the global imprint of INTECHOPEN LIMITED, registered in England and Wales, registration number: 11086078, 167-169 Great Portland Street, London, W1W 5PF, United Kingdom

For EU product safety concerns: IN TECH d.o.o., Prolaz Marije Krucifikse Kozulić 3, 51000 Rijeka, Croatia, info@intechopen.com or visit our website at intechopen.com.

British Library Cataloguing-in-Publication Data
A catalogue record for this book is available from the British Library

Cellulose - Biobased Solutions for Society
Edited by Alejandro Rodríguez
p. cm.

This title is part of the Materials Science Book Series, Volume 8
Topic: Composite Materials
Series Editor: Chonghe Li
Topic Editor: Anna Boczkowska
Associate Topic Editors: Kamil Dydek and Bartlomiej Przybyszewski

Print ISBN 978-1-83634-533-6
Online ISBN 978-1-83634-532-9
eBook (PDF) ISBN 978-1-83634-534-3
ISSN 3049-8856

IntechOpen Book Series

Materials Science

Volume 8

Aims and Scope of the Series

Materials science has always occupied an extremely high position in the human development process. As we explore the oceans of stars, various industries have put forward more stringent requirements for the performance of materials, forcing us to pay more and more attention to the development of new materials. At the same time, the formation of a data-driven scientific paradigm is dramatically shortening the development cycle of new materials. The huge data generated by synergistically combining theories, high-throughput experiments, high-throughput computation, and artificial intelligence is greatly contributing to our ability to utilize materials science to solve real-world problems. The three topics of this book series - Metals and Nonmetals; Composite Materials; and Surface Science - will address important areas of advancement in materials science. There will be a range of interesting works published under these topics.

Meet the Series Editor

Prof. Chonghe Li received his Ph.D. from the Chinese Academy of Sciences in 1995. From 1995 to 2000, he worked as a researcher at the Shanghai Institute of Metallurgy, Chinese Academy of Sciences, where he also served as director of the research laboratory. In 2000, he was appointed Professor at the Institute of High-Performance Computing in Singapore, where he worked on computation and simulation in materials science until 2004. Since then, he has been a professor at the School of Materials Science and Engineering, Shanghai University, China, as well as the director of the Shanghai Specialty Casting Engineering and Technology Research Center. Prof. Li's research focuses on titanium alloys, titanium-aluminum single crystals, intermetallic compounds, theoretical calculations, alloy design, and special refractory materials. His broad scientific expertise is well recognized by the scientific community around the world. He is a member of the editorial board of the journal Metals. As an author, he has published more than 200 peer-reviewed papers, 2 books, and over 40 patents.

Meet the Volume Editor

Dr. Rodríguez is a Full Professor in the Chemical Engineering Area at the Universidad de Córdoba. His research focuses on the valorization of various lignocellulosic material fractions through environmentally friendly biorefinery processes. He has published 140 indexed articles and participated in over 100 international congresses. He has supervised 11 doctoral theses and is currently supervising four. Dr. Rodríguez collaborates with several national and international research groups. He is a member of the Ibero-American Network for Research on Pulp and Paper. He coordinates a research group involved in Provalor and Nanocelia, both of which are part of the CYTED thematic networks. He also serves on scientific committees for multiple international conferences. He has chaired the Organizing Committee for all three editions of the International Workshop on Biorefinery of Lignocellulosic Materials. He is an expert evaluator for numerous international agencies, including the European Research Agency (REA), ANEP (Spain), Fondazione Cariplo (Italy), Tec21 (France), SGS, EQA, DNVGL, ACIE (Spain), the Executive Agency for Higher Education, Research, Development and Innovation Funding (Romania), Luxembourg National Research Fund, CONICYT (Chile), Latvian Council of Science, the Knowledge Foundation (Sweden), and the Pontifical Catholic University of Peru, among others. Dr. Rodríguez is the Head of the Bioprocess and Products Engineering research group (RNM 940), the Academic Director of the Master's Degree in Chemical Tools for the Agri-food and Environmental Industry, and the General Director of OTC within the Vice-Rectorate for Innovation and Transfer at the University of Córdoba.

Contents

Preface **XIII**

Section 1
Current Innovations in Recovering Cellulose **1**

Chapter 1 **3**
Recovery of Cellulose from Food and Agricultural Waste
by Symone Alexander and Mariama Alidu

Section 2
Nanocellulose **27**

Chapter 2 **29**
Bacterial Nanocellulose as a Green Material for Food Packaging Application
by Bhargavi Pathak, Sukanya Saikia, Anurag Kashyap, Runjun Sarma, Munmi Sarma and Shahida Choudhury

Chapter 3 **49**
Nanocellulose, a Promising Raw Material: Improving the Scalability of TEMPO-Mediated Oxidation Process
by Jose Luis Sanchez-Salvador, Hongyu Xu, Ana Balea, Elena Fuente, M. Concepcion Monte, Angeles Blanco and Carlos Negro

Chapter 4 **67**
Nanocellulose for Wastewater Treatment
by Madhu Sasikala, K. Sivaprakash, M. Lavanya and V. Ramesh Prabu

Section 3
Functional Materials Obtained from Agri-Food Residues **81**

Chapter 5 **83**
Lignocellulosic Biomass to Produce Functional Materials via DES: Environmental Remediation Applications
by Remedios Montenegro, Zoilo González and Alejandro Rodríguez Pascual

Chapter 6 **107**
Biobased Composites: A Sustainable Pathway for Modern Applications
by Behrooz Maleki, Pouya Ghamari Kargar and Samaneh Sedigh Ashrafi

Preface

The agri-food industry is becoming increasingly important, among other reasons, because the world population is growing at a significant rate and is expected to reach 10 billion inhabitants by 2050. Every year, this industry generates large quantities of residues, known as plant biomass or lignocellulosic biomass, which have a high potential for recovery due to their composition, including cellulose, hemicelluloses, lignin, bioactive compounds, and polyphenols, among others. The broad spectrum of applications of these residues in sectors as diverse as food packaging, biocomposite formulation, biofuel generation, and environmental remediation represents an excellent market opportunity to improve an essential sector for the global population, namely agriculture. Thus, recovering this residue represents an environmental, social and economic benefit that can undoubtedly bring society closer to real models of bioeconomy and sustainable development.

Chapter 1, "Recovery of Cellulose from Food and Agricultural Waste", by Symone Alexander and Mariama Alidu, covers current innovations in recovering cellulose from food and agricultural waste, focusing on recovery and key characterization techniques for cellulose extracted from food and agricultural waste. It also describes the challenges associated with various waste feedstocks and the processes used to extract cellulose and outlines potential solutions.

In Chapter 2, "Bacterial Nanocellulose as a Green Material for Food Packaging Application" by Bhargavi Pathak et al. describe the synthesis and bioengineering of harnessing BNC to create green, active, and smart solutions addressing contemporary challenges in the food packaging industry.

In Chapter 3, "Nanocellulose, a Promising Raw Material: Improving the Scalability of TEMPO-Mediated Oxidation Process", by Jose Luis Sanchez-Salvador et al., several strategies to enhance the CNF production and optimize the overall process are reviewed to show the significant and critical progress carried out in the last decade towards achieving more efficient and sustainable nanocellulose production.

Chapter 4, "Nanocellulose for Wastewater Treatment" by Sasikala M et al., highlights recent advancements in nanocellulose applications, including the removal of heavy metals, separation of oily water, and extraction of dyes, showcasing its potential as an effective adsorbent for improving filtration performance.

Chapter 5, "Lignocellulosic Biomass to Produce Functional Materials via DES: Environmental Remediation Applications" by Remedios Montenegro et al., addresses the use of diverse types of DES and their impact on the structural and chemical composition of various (ligno)cellulosic raw materials. Additionally, it explores the potential applications of cellulose-conformed materials, with an emphasis on nanoscale materials, in remediation processes, highlighting the enhanced outcomes achieved through the utilization of DES.

Chapter 6, "Biobased Composites: A Sustainable Pathway for Modern Applications" by Behrooz Maleki et al., describes the integration of agricultural waste, such as hemp, flax, and jute, with biopolymers like polylactic acid (PLA) and polyhydroxyalkanoates (PHA) fosters advancements in material performance, including improved strength, stiffness, and thermal properties.

I want to thank the management at IntechOpen for their support during the editing of this book.

Alejandro Rodríguez Pascual
Full Professor,
Inorganic Chemistry and Chemical Engineering Department,
Institute of Chemistry for Energy and the Environment (IQUEMA),
University of Córdoba,
Córdoba, Spain

Section 1

Current Innovations in Recovering Cellulose

Chapter 1

Recovery of Cellulose from Food and Agricultural Waste

Symone Alexander and Mariama Alidu

Abstract

This chapter covers current innovations in recovering cellulose from food and agricultural waste. The production of waste is increasing daily, which has led to a growing emphasis on cellulose-based materials in the literature as of 2024. Extracting high-value products like cellulose from waste resources can significantly reduce emissions that are detrimental to the environment. This chapter focuses on recovery and key characterization techniques for cellulose extracted from food and agricultural waste. It also describes the challenges associated with various waste feedstocks, the processes used to extract cellulose, and outlines potential solutions. Additionally, we will discuss opportunities for innovation and advancement along with the current state of the art.

Keywords: cellulose, biomass, waste valorization, upcycling, food waste, agricultural waste

1. Introduction

Food waste refers to edible food discarded without being consumed by humans. According to Sustainable Development Goal Target 12.3 of the Environmental Protection Agency (EPA, USA), food waste can originate from the retail, food service, and household levels (**Figure 1**) and is often discarded through landfilling, composting, or littering [1–3]. The environmental consequences of food waste cannot be overstated. When food is discarded in landfills, it generates methane—a greenhouse gas that is up to 36× more environmentally potent than carbon dioxide [4]. The United States Food and Agriculture Organization (FAO) reports that food waste accounts for approximately 8% of global greenhouse gas emissions. In the United States, organic waste is the single most significant contributor to methane emissions, releasing toxic gases equivalent to the annual emissions of 37 million cars [1].

Agricultural waste refers to unwanted materials generated exclusively from agricultural operations directly associated with commercial crop production. This category includes waste from vegetable and fruit-bearing plants and trees but excludes landscape vegetation such as weeds, grass, and tree trimmings [5]. Agricultural waste is generated at a rapid rate, with an average annual increase of 5–10%, leading to

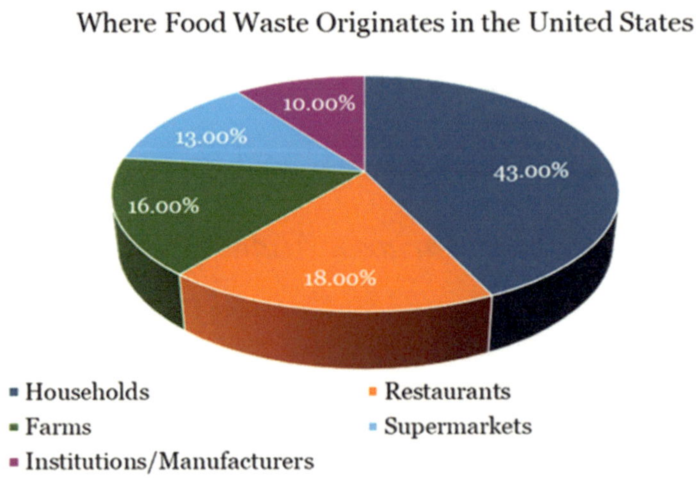

Figure 1.
Sources of food waste in the United States adapted from [1].

significant environmental concerns. The practice of burning these wastes contributes to water, soil, and air pollution [5].

Food and agricultural waste contain valuable components that can be extracted and utilized to produce various materials. One such component is cellulose, a polysaccharide composed of linear (1,4)-linked β-D-glucan chains, each with three reactive hydroxyl groups per anhydroglucopyranose unit [6]. Due to its highly crystalline structure, excellent biodegradability, abundant availability, favorable mechanical properties, and outstanding biocompatibility, cellulose has garnered significant attention from researchers to develop a wide range of functional materials [6].

Cellulose is structurally organized into fibrils, which are encased in a matrix of lignin and hemicellulose. The isolation of cellulose is typically achieved through a rigorous chemical process known as pulping. During pulping, lignin and hemicellulose are fragmented by breaking their ether linkages and are then solubilized in an appropriate medium [7]. This process enables the extraction of cellulose in its native form, cellulose I. The crystallinity of native cellulose consists of two distinct crystalline phases, cellulose Iα and Iβ, with their ratio varying depending on the source of cellulose. Cellulose Iα has a triclinic crystal structure, while cellulose Iβ has a monoclinic structure. In addition to these native phases, cellulose can also exist in several other crystalline forms, such as cellulose II, III_I, III_{II}, IV_I, and IV_{II}, which can be synthesized through thermochemical treatment of cellulose [8].

The different types of nanocellulose are typically classified into cellulose nanocrystals (CNC) or cellulose nanofibrils (CNF) based on factors such as their dimensions, functions, preparation methods, and production source. CNCs are primarily derived from cellulose through acid hydrolysis and are known for their low density, high specific surface area, and high crystallinity. CNCs possess numerous surface hydroxyl groups, enabling various chemical modifications, such as oxidation and esterification [9, 10]. CNFs are synthesized through mechanical processes such as homogenization, grinding, and ball milling. These methods produce flexible, fibril-shaped particles forming long fiber networks with diameters larger than CNCs. Fibrils with smaller diameters and greater lengths exhibit enhanced reinforcing effects in composite materials [11]. CNCs have diameters ranging from 2 to 70 nm and lengths ranging from 100 to 600 nm, while CNFs have diameters ranging from 5 to 30 nm and lengths greater than 1 μm [12].

1.1 Sources of extracted cellulose

The primary source of cellulose is natural fibers. In 1833, Anselme Payen conducted an extensive analysis of wood and discovered cellulose [13]. Cellulose can be obtained from a variety of sources, including plants [14], algae [15], bacteria [16], and tunicates [17]. However, plants remain the most significant source of cellulose production due to their abundance and cost-effectiveness. Cellulose is also the primary component of the cell wall in many algal species, with green algae being the most suitable source for cellulose extraction. The bacterium Komagataeibacter xylinus uses carbon and nitrogen sources to produce thick, clear cellulose microfibrils in the growth medium. Additionally, enzyme complexes found in the epidermal layer of tunicates are responsible for cellulose production [18].

1.2 Sources of cellulose from waste

1.2.1 Forestry

The United States has the fourth-largest forest estate in the world, comprising approximately 8% of the global forest area [19]. With about 3 million square kilometers of forest land, these forests cover roughly a third of the country's total land area [19]. The forests span a variety of ecosystems, including boreal forests in Alaska, deciduous forests in the eastern United States, pine forests in the southern United States, dry coniferous forests in the western United States, temperate rainforests on the West Coast, and tropical rainforests of Puerto Rico and Hawaii [19]. US forests are steadily growing, and the potential for producing forest residues—such as small-diameter trees produced during harvest operations—is considerable.

The valorization of forest residues has become a key focus in green chemistry and the circular bioeconomy, offering significant opportunities to produce high-value materials. However, the efficient extraction of cellulose from forest residues presents substantial technical challenges, primarily due to the complex structure of the plant cell wall and the presence of lignin and hemicelluloses [20]. Additionally, deforestation poses a significant long-term threat to the environment's health.

1.2.2 Food and agriculture waste

Food and agricultural waste contribute significantly to the waste generated in the United States. Approximately one-third of the food produced worldwide is wasted, totaling around 1.3 billion tons annually [2]. In Europe and North America, consumer food waste per capita is estimated to be 95–115 kg/year [2]. The uncontrolled disposal of biodegradable waste release greenhouse gases, which have harmful effects on human health [21]. About 45% of food waste consists of fruits and vegetables, which can be transformed into viable feedstocks for producing useful biopolymers [22]. Food and agricultural waste typically poses minimal harm to the environment in small quantities. However, when accumulated in large volumes, it can harm humans, animals, and plants [23]. Food and vegetable waste is abundant and has a lower lignin content, ranging from 10 to 25%, compared to forestry biomass, which typically has a lignin content of 20–30% [24]. This lower lignin content simplifies the cellulose extraction process. However, the conversion and valorization of food waste to produce cellulose is relatively new, and standardization and quality control are needed to mitigate potential economic, environmental, and social risks [25].

2. Extraction process overview

Extraction is defined as the process of transferring a compound from its original state into another phase. It can also be described as the separation of the compound from its surrounding matrix [26]. Several techniques are employed to extract cellulose from a matrix, including milling, boiling, acid treatment, alkaline treatment, bleaching, dialysis, and others, depending on the desired end product of the cellulosic material. In this section, we provide a general overview of the steps used to extract cellulose from food and agricultural waste.

2.1 Size reduction

Size reduction is typically the first step in the extraction process and is the application of high mechanical force to break the rigid structure of lignocellulose. Size reduction is typically achieved through methods such as milling, grinding, ultrasonication, crushing, and stirring. The goal is to homogenize the waste stream and to reduce particle size, degree of polymerization, and lignin content. Mechanical treatment also improves the surface area needed for chemical processing. The advantages of size reduction include simplicity and ease of processing, reduced crystallinity, increased porosity, increased enzyme digestibility, and low initial investment. However, the disadvantages include high operational costs, high energy consumption, the inability to completely degrade lignin, and the need for combination with other pretreatments [5].

2.2 Boiling

During extraction, boiling is used to remove non-cellulosic components such as hemicellulose and to break down the cell wall [22]. Due to the numerous water-soluble components (WSC) in food and agricultural waste, the boiling step is usually repeated 2–3 times and can range in time from 10 to 60 min. As an alternative, steam explosion has been employed to remove hemicellulose and other WSC in lignocellulosic materials. Steam explosion technology is used for the pretreatment of food waste due to its efficiency, low cost, and environmental benefits. These processes are typically initiated at temperatures ranging from 160 to 260°C (corresponding pressure of 0.69–4.83 MPa) and can last from several seconds to a few minutes. The food or agricultural waste/steam mixture is held at the selected temperature and pressure needed to promote the hydrolysis of hemicellulose, and then the mixture undergoes explosive decompression to obtain a slurry [27, 28]. This process causes hemicellulose and lignin to degrade, enhancing the potential for cellulose recovery [28]. However, research on this technology is still limited. Additionally, the impact of steam explosion on the physicochemical properties and functional components of raw materials requires further investigation [29]. Typically, after boiling or steam explosion, the mixture is filtered to separate the insoluble pulp from the filtrate containing WSC and a residue containing pectin, remaining hemicellulose, and lignin [22].

2.3 Acid treatment

Acid treatment hydrolyzes the residual hemicelluloses and pectin and breaks them down into simple sugars. After the degradation of the amorphous domains, individual lignocellulosic crystallites and fibers are liberated, achieving nano to micro-scale dimensions [30]. As the less ordered domains are reduced, an increase in cellulose content and crystallinity is expected, which is proportional to the reduction of the amorphous regions. **Table 1** summarizes the benefits and disadvantages of the acids used in cellulose extraction, focusing on cost, safety, and environmental impact.

2.4 Alkaline treatment

Alkaline reagents, such as sodium hydroxide (NaOH) and potassium hydroxide (KOH), are used to depolymerize and liquefy lignin and hemicellulose, reduce crystallinity in cellulose and enhance pretreatment efficiency, ultimately increasing extraction rates. Elevated temperatures (75–125°C) are used in combination with NaOH to assist in depolymerizing lignin, reducing the crystallinity of lignocellulosic

Acid	Benefits	Disadvantages
Sulfuric acid (H_2SO_4)	Sulfuric acid treatment hydrolyzes almost all of the hemicellulose content	*Cost per 100 ml*: $139 (35)
		Safety: Reacts violently with water and has an NFPA health rating of 3.
		Environmental impact: Exhibits moderate acute toxicity to aquatic life
Hydrochloric acid (HCL)	HCl, a milder acid,	*Cost per 100 ml*: $52.40
		Safety: Should not be released into the environment, as HCl gas may be emitted. NFPA health rating of 3
		Environmental impact: Hazardous vapors that could contribute to acid rain.
Nitric acid (HNO_3)	Higher crystallinity index and cellulose content compared to H_2SO_4	*Cost per 100 ml*: $57.40
		Safety: Highly reactive and a significant explosion hazard, with an NFPA health rating of 4.
		Environmental impact: Releases nitrogen oxides (NO_x), contributing to air pollution.
Phosphoric acid (H_3PO_4)	Greater thermal stability compared to H_2SO_4	*Cost per 100 ml*: $45.50
		Safety: Can react violently with bases. NFPA health rating of 3.
		Environmental impact: Excessive use may contribute to eutrophication in aquatic environments.
Perchloric acid ($HClO_4$)	Increases hydrophobicity and thermal stability and provides better control over the size of CNCs.	*Cost per 100 ml*: $93.90
		Safety: Toxic, corrosive, and possesses oxidizing properties. NFPA health rating of 3, instability rating of 3, potential for explosive decomposition *Environmental impact*: Mist and vapor can condense in ventilation systems, forming highly explosive metallic perchlorates.

Table 1.
Acids used to hydrolyze hemicellulose and lignin during cellulose extraction.

feedstock, and increasing the contact surface of cellulose [5]. Sodium hydroxide can also be used to modify the cellulose type post-extraction. For example, the treatment of cellulose I with aqueous sodium hydroxide (16.5% by weight) resulted in the formation of cellulose II, a much less crystalline form of cellulose [31]. Romruen et al. extracted cellulose from rice straw, corn cob, and pineapple leaf using 5% KOH; the treatment resulted in higher cellulose content by removing non-cellulosic components [32]. In the treatment of eucalyptus, kraft pulp, and beech sulfite using 10% NaOH and 14% KOH, higher cellulose content was achieved with NaOH for eucalyptus pulp, while lower efficiency was observed for beech sulfite pulp. These findings suggest that the efficiency of KOH and NaOH in cellulose extraction can vary depending on the type of biomass [33].

2.5 Bleaching

Lignin is known to contribute to the brown coloration of various plant materials. The final step in the cellulose extraction process is typically to remove residual lignin through a bleaching process. Various compounds, such as hydrogen peroxide (H_2O_2), are used as bleaching agents to break the different types of bonds within lignin and to lighten the fiber's color to a white or neutral shade. The bleaching process can affect the physical properties of the fiber, including its strength and durability [22, 34–39]. Additionally, the use of sodium chlorite and hydrogen peroxide, due to their oxidizing properties, plays a key role in the final quality of cellulose fibers and the remaining lignin content [22]. The decomposition of H_2O_2 into hydroxide (OH^-) and oxygen (O_2) helps to break down the lignin structure and prevents the release of inhibitory side products [5]. H_2O_2 is a milder bleaching agent compared to sodium chlorite, making it more suitable for high-yield pulp, and its reaction products are relatively non-toxic. However, H_2O_2 is prone to degradation [40]. On the other hand, $NaClO_2$ is highly effective at delignifying cellulose [41, 42]. However, it is non-selectively toxic in drinking water or aquatic environments, making it much less environmentally friendly [43].

2.6 Dialysis

Dialysis is often used to purify the extracted cellulose, particularly CNCs and CNFs. In the dialysis method, the precipitates obtained from the extraction process are re-suspended in water and subjected to dialysis using regenerated cellulose dialysis membranes (or tubes) with a molecular weight cut-off of 12–14 kDa. This method removes non-reactive functional groups (i.e., sulfates) and eliminates different salts and soluble sugars. Depending on the experiment, dialysis may take from several days to 2 weeks to achieve a constant pH. Neutralization has been proposed as an alternative method to reduce production time for CNC preparation [30].

3. Cellulose characterization

3.1 Initial cellulose content

The evaluation of the initial cellulose content in biomass feedstocks typically begins with a literature review followed by experimental methods. For example, the initial cellulose content of cotton fibers is well-established, whereas, for unique species, such as the extraction from three different timber species [20], the initial cellulose content may be novel and require experimentation. **Table 2** summarizes the initial cellulose content values obtained both experimentally and through literature research.

The yield can then be calculated based on the determination of the initial cellulose content Eq. (1) or based on the overall mass of the feedstock Eq. (2).

$$\text{Yield} = \frac{\text{Weight of dried extracted cellulose}}{\text{Weight of initial cellulose content}} \tag{1}$$

$$\text{Yield} = \frac{\text{Weight of dried extracted cellulose}}{\text{Weight of dried agriculture / food waste}} \tag{2}$$

Table 2 provides the yield data from the selected food and agricultural waste feedstocks.

3.2 Techniques for characterization

Advancements in the broad field of analytical techniques have enabled the characterization of cellulose to determine its mechanical properties, morphology, crystallinity, and thermal properties [18]. Several techniques, both individually and in combination, have been employed to extract valuable structural details about the components of the biomass feedstock and the resultant cellulose [51]. This section details the key methods for characterizing the structure and interactions of extracted cellulose.

3.2.1 Molecular weight and degree of polymerization

The number of monomer units in cellulose and in polymers is referred to as the degree of polymerization (DP) [52]. The degree of polymerization for extracted cellulose is closely related to the aspect ratio of the nanofibers, and any pretreatment method tends to decrease the degree of polymerization [18]. The most common methods for determining the molecular weight of cellulose are viscometry

Feedstock	Initial cellulose content	Method	Extracted cellulose yield
Navel Orange [44]	85.80%	ASTM D1103-60	26.20%
Tea leaf fiber [45]	16.20%	Experimental	87.9%
Corn cob [46]	31.2 ± 3.1%	Acid chlorite method and KOH [47]	46–57%
Corn stover [48]	45.50%	m TAPPI T19-54	64%
Rice straw [37]	38.30%	NREL/TP-510-42,618	36%
Sugarcane straw [49]	32-45%	Van Soest procedure (AOAC) [50]	22.7 ± 2.5%

Table 2.
Initial cellulose content and yield of selected food and agricultural waste.

and size exclusion chromatography (SEC), often coupled with light scattering and refractive index detectors [22, 52, 53]. Interestingly, not many references determine the molecular weight of the cellulose extracted from food and agricultural waste. We include the limited number of references in (**Table 3**) and discuss their procedures below.

The TAPPI T230-99 protocol is commonly used in combination with a viscometer to determine the molecular weight and degree of polymerization of cellulose [54]. The sample consists of approximately 0.250 g of cellulose dissolved in a 0.5 M cupriethylenediamine solution. The viscometer then measures the time it takes for the solution to move between two designated points. The intrinsic viscosity is calculated from the experimental measurements using the Mark-Houwink equation and then is used to calculate the molecular weight and degree of polymerization of the cellulose Eqs. (3) and (4) [22].

$$[\eta] = 3.85 \times 10^{-2} (MW)^{0.76} \tag{3}$$

$$[\eta] = 1.75 (DP)^{0.69} \tag{4}$$

SEC, also known as gel permeation chromatography (GPC), is a specialized type of column chromatography in which it is assumed that no enthalpic interactions occur between the analyte and the stationary phase. Molecules are separated based solely on their hydrodynamic volume, with larger molecules (in terms of hydrodynamic volume) eluting first. Linear polysaccharide pullulan is often used for calibration due to the lack of commercially available cellulose standards [52]. The universal calibration is valid only if the separation process is based purely on entropic effects. Therefore, if there are other interactions between the polymer and the stationary phase, universal calibration becomes invalid. Universal calibration employs the Einstein viscosity law Eq. (5), which demonstrates that the product [η] × M is a direct measure of the molecule's hydrodynamic volume. For cellulose in SEC, N, N-dimethylacetamide/lithium chloride (DMAc/LiCl) is frequently used as the solvent and eluent [52].

$$[\eta] = K \left(\frac{V}{M} \right) \tag{5}$$

3.2.2 Crystallinity

Crystallinity is significantly influenced by the lignocellulosic source [18]. Generally, cellulose nanocrystals derived from H_2SO_4 show lower crystallinity values compared to those made from HCl. Additionally, increasing hydrolysis time improves

Feedstock	Method	MW or [η] of extracted cellulose	Reference
Mixed waste	TAPPI T230-99	MW ~1323 Da	[22]
Navel oranges	ASTM D1103-60	[η]- 403 ml/g	[44]

Table 3.
Methods for determining the molecular weight of extracted cellulose.

crystallinity by removing amorphous regions. Crystallinity is one of the most important characteristics of cellulose, contributing to its physical, chemical, and mechanical properties [55]. The most widely used method to determine the crystallinity index (CrI) is X-ray diffraction (XRD) [55]. Segal et al. proposed a method to determine the relative crystallinity of cellulose samples using XRD [56]. The Segal method Eq. (6) is based on the premise that cellulosic material consists of two components: non-crystalline (amorphous) and crystalline. The amount of amorphous material is represented by the minimum intensity between the major peaks, while the amount of crystalline material is represented by the height of the highest diffraction peak [57].

$$CrI = \frac{I_{002} - I_{AM}}{I_{AM}} \tag{6}$$

Where I_{002} is the Miller indices at 22.5°, and I_{AM} is the intensity of the amorphous region at 18°.

Cellulose is organized into small crystals, and there are also several different crystal forms (polymorphs) that depend on the history of the sample [57, 58]. French described the various patterns in cellulose types I (α, β), II, III_I, and III_{II} [58]. **Table 4** displays the cellulose type and percent crystallinity for cellulose extracted from food and agricultural waste in the literature.

3.2.3 Thermal stability

Several techniques, such as thermogravimetric analysis (TGA), differential scanning calorimetry (DSC), and thermomechanical analysis (TMA), can be employed to assess the thermal stability of cellulose. TGA measures the mass changes in cellulose, DSC evaluates the glass transition, melting, and crystallization, and TMA determines thermal expansion and shrinkage [51]. Among these methods, TGA is the most commonly used for determining the thermal stability of extracted cellulose [63].

TGA analysis is often conducted at a heating rate of 10°C/min under a nitrogen atmosphere, with nitrogen gas flow rates ranging from 5 to 100 mL/min considered acceptable [63]. The onset temperature of thermal decomposition (T_{onset}) is determined by tracing two tangents on the thermogram. According to

Feedstock	% Crystallinity/type	Reference
Mixed waste	16%/Type I	[22]
Navel oranges	68.7%	[44]
Tomatoes	80.8%	[59]
Tea leaves	83.1%	[45]
Pineapple leaves	49–73%/Type 1	[60]
Corn cobs	83.7%/Type 1	[46]
Corn stover	50.04%	[48]
Rice straw	91.2%	[61]
Kenaf fibers	81.8%	[62]

Table 4.
Crystallinity of extracted cellulose from various feedstocks.

ASTM E2550-07 (2012), T_{onset} is defined as the point where the TGA curve starts to deviate from the tangent line after dehydration [63]. Thermal degradation of polymeric materials like cellulose typically involves two main stages: the elimination of volatile matter or water evaporation and the depolymerization process [64]. The thermal degradation of lignocellulosic materials begins with the initial decomposition of hemicelluloses, followed by the pyrolysis of lignin, depolymerization, flaming combustion, and char oxidation. Additionally, cellulose nanofibrils exhibit a high degradation temperature onset (around 350°C) and demonstrate superior thermal performance compared to hemicellulose, pectin, and lignin. In contrast, the thermal degradation of cellulose nanocrystals typically begins between 200 and 300°C. Cellulose nanocrystals with a lower sulfate content tend to exhibit better thermal stability [18].

3.2.4 Fiber properties

The morphology, chemical functional groups, and mechanical properties of cellulose can be analyzed using various techniques such as scanning electron microscopy (SEM), Fourier transform infrared (FTIR) spectroscopy, and dynamic mechanical analysis (DMA). The morphology of cellulose varies depending on the extraction process. The mechanical properties of cellulose nanocrystals, such as elastic modulus, range from 110 to 220 GPa, while their tensile strength ranges from 7.5 to 7.7 GPa [18]. FTIR spectroscopy is commonly used to assess the overall purity of the extracted cellulose material. Salim et al. used FTIR to study the formation of noncovalent bonds in cellulose [65]. These properties significantly influence the physical characteristics of cellulose, including its solubility. FTIR analysis is typically conducted within the wavenumber range of 4000–400 cm^{-1} (**Table 5**).

Peak wavenumber (cm^{-1})	**Functional group**	
582	C–OH	Carboxyl
655	C–OH	Carboxyl
891	C–H	Methyl
1017	C–O	Carbonyl
1160	C–O–C	Ethers
1317	C–O	Carbonyl
1376	C–H	Alkane
1437	CH_2	Alkane
1635	H–O–H	Absorbed water
2019	C=C	Alkene
2129	C–C	Alkyne
2350	–COO	Carbon dioxide
2899	C–H	Methyl group
3318	OH	Hydroxyl group for α-cellulose

Table 5.
Summary of the peak and IR bands of cellulose adapted from [65].

4. Feedstocks

Over the past 5 years, a wealth of different food and agricultural waste feedstocks for cellulose have been identified. In this section, we highlight the latest innovations and provide flowcharts of the general extraction process for each feedstock.

4.1 Sugarcane

Sugarcane straw and bagasse are by-products of the sugar industry. They consist of approximately 62% dry leaves, 32% green leaves, and 7% [66]. According to the UN FAO, global sugarcane production reached 1.9 billion tons in 2021 [67]. Often, sugarcane by-products are burned in open landfills, contributing to air and water pollution [68]. Due to their structural composition, which is high in cellulose content, sugarcane straw holds significant potential as a lignocellulosic biomass [49]. The extraction process involves pretreatment with boiling water and ethanol, followed by bleaching and treatment with alkali to remove residual lignin (**Figure 2**).

4.2 Corn

Corn is one of the most widely produced crops in the United States, with a production volume of 350 million metric tons in the 2023/2024 season. The states of Iowa, Nebraska, and Illinois account for over 30% of this cultivation. This production volume represents approximately one-third of the global corn output [69]. From a utilization perspective, corn serves as a staple food in many parts of the world, is extensively used as animal feed, and has recently gained significance as an industrial feedstock. It is increasingly employed in the production of biofuels, chemical compounds, bioplastics, and other materials [70]. Despite its significance, many components of the maize plant, such as corn cobs and corn stover, remain underutilized. Corn stover is one of the most abundantly produced agricultural residues in the United States and is comprised of 34.5% stems, 32.3% leaves, 14.3% husks, and 12.3% cobs [71]. Its chemical composition includes approximately up to 45% cellulose, up to 39% hemicellulose, and up to 20% lignin [71].

Corncob is a residual by-product generated after maize harvesting and grain removal, performed either mechanically or manually. On average, corncobs account for approximately 14% of the grain yield and represent about 16% of the total corn stover in a field. Unlike other components of stover, corncob waste has minimal effects on soil fertility due to its low biodegradability. Additionally, its removal from crop fields has been shown to potentially increase maize yield [72]. The potential of corncobs for high-purity cellulose extraction is significant, owing

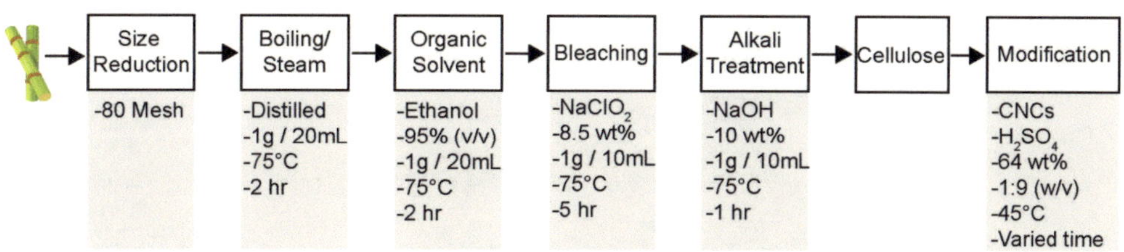

Figure 2.
Overview of the cellulose extraction process from sugarcane feedstocks and the modification to form cellulose nanocrystals.

to their high cellulose content (34–40%) and hemicellulose content (36–41%), combined with a relatively low lignin content (9–19%) [73]. Traditionally, corncobs are primarily used for producing animal feed and fertilizers. However, in industrial applications, corncobs serve as a valuable raw material to produce bioenergies such as bioethanol, enzymes like cellulases, and by-products such as xylose and glucose [46, 72].

Corn stover refers to the above-ground portion of the maize plant that remains after grain harvesting, accounting for approximately 47–50% of the total grain yield on a dry mass basis [71]. Currently, only about 5% of corn stover is utilized as cattle feed, while a small fraction is left in the field to enhance soil fertility. This leaves a significant amount of stover unused. Its composition, particularly the high cellulose and hemicellulose content, makes corn stover a promising feedstock for cellulose nanofiber and nanocrystal production [23]. Additionally, the entire corn stover can be converted into various high-value end products, as illustrated in **Figure 3**.

The extraction processes for corn cobs and corn stover are fairly identical, with minor changes in temperature for the alkali, bleaching, and CNC production steps. Corn cobs require much less bleaching time than corn stover (5 h vs. 24 h). However, the modification of cellulose to cellulose nanocrystals only required 20 minutes for corn stover relative to 30–90 min for corn cobs, with only a 5-degree difference in temperature.

4.3 Rice straw

Rice is the most consumed grain in the world, particularly in countries such as China, Bangladesh, and India. In 2023, rice production in the United States reached 11 million metric tons, with Arkansas, California, and Louisiana contributing approximately 5.4 million, 2.2 million, and 1.6 million metric tons, respectively [74]. Every 1 kg of rice harvested is accompanied by 1–1.5 kg of rice straw, while an optimal milling process can yield around 20% rice husk.

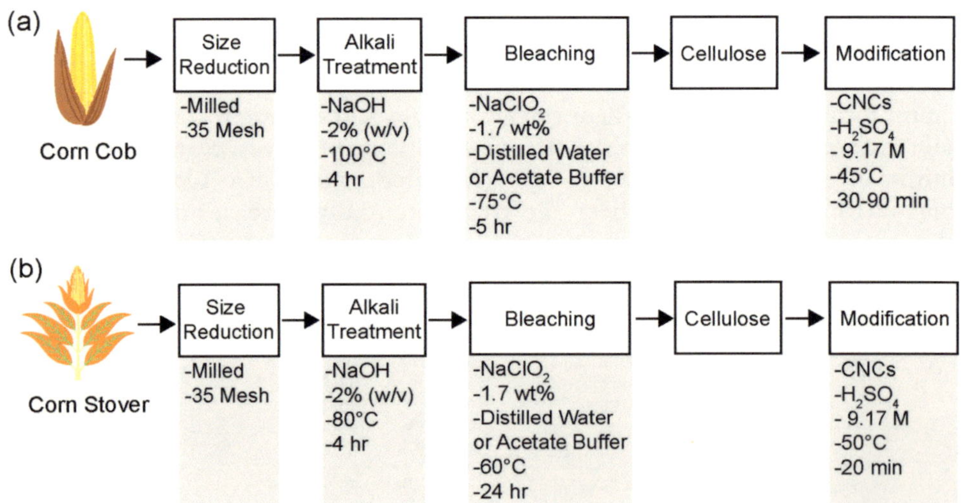

Figure 3.
Overview of the cellulose extraction process from corn cob and corn stover feedstocks and the modification to form cellulose nanocrystals.

Traditionally, rice straw has been utilized as animal feed, feedstock for the paper industry, and as an organic fertilizer. The lignocellulosic composition of rice straw consists of 32–47% cellulose, 19–27% hemicellulose, and 5–24% lignin [75]. Rice straw is considered the most abundant cellulose source among agricultural crop residues worldwide. Lu further described methods for extracting high-quality cellulose from rice straw to produce cellulose nanocrystals [61]. The major addition to the standard extraction process is the use of a toluene/ethanol organosolv pretreatment step, which was used to remove non-cellulosic extractives (**Figure 4**).

4.4 Pineapple leaf fibers

Pineapples are one of the most widely recognized tropical fruits and have multiple components that are rich in cellulose. The pineapple processing industry is projected to generate approximately 50–60% solid waste by 2024 [76, 77]. This waste, which includes roots, stems, and leaves, is typically disposed of through methods such as burning or used in low-value applications like animal feed [60, 78]. Among these waste materials, pineapple leaf fiber (PALF) stands out due to its resistance to degradation, attributed to its high cellulose and lignin content [36].

PALF is composed of 70–85% cellulose, 6–19% hemicellulose, 4–15% lignin, less than 4% wax, and 1–5% ash. This composition makes PALF an excellent raw material for the extraction of cellulose microfibers, nanofibers, and nanocrystals and a promising material for reinforcement in composite matrices [36, 60]. The extraction process for cellulose from pineapple waste follows the typical process of alkali pretreatment followed by bleaching. It does not require any organic solvent, which

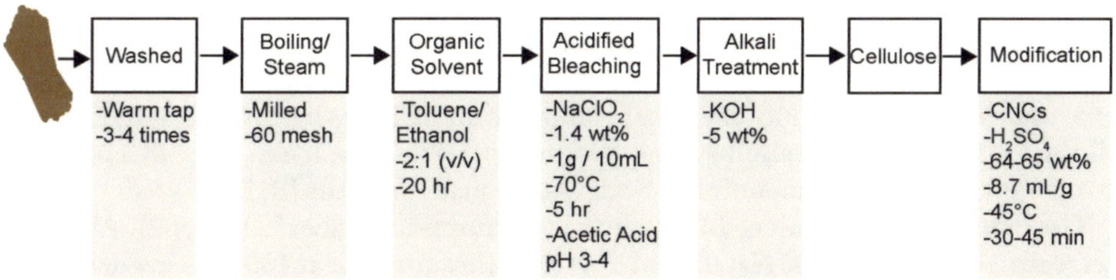

Figure 4.
Overview of the cellulose extraction process from rice waste feedstocks and the modification to form cellulose nanocrystals.

is advantageous from an environmental perspective, but current processes still use a chlorinated bleaching agent (**Figure 5**).

4.5 Tea leaf fibers

Tea leaves are consumed in large quantities globally, generating significant amounts of waste [79]. An estimated 78% of tea production consists of black tea, while 20% is green tea, and by 2027, tea waste is projected to reach 7.2 million tons [79]. Tea residue is rich in polysaccharides, including cellulose, hemicellulose, lignin, caffeine, and other compounds. Specifically, tea leaf waste fiber contains approximately 16.2% cellulose, 68.2% hemicellulose, and 18.8% lignin [79]. The negative environmental impacts of disposing of tea waste can be mitigated by exploring its potential applications in industries as an

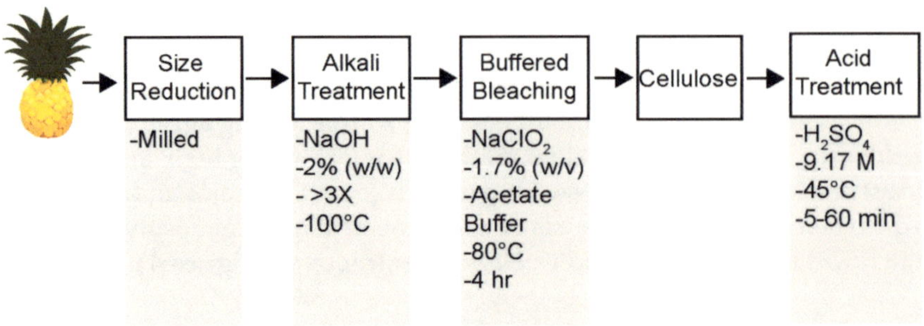

Figure 5.
Overview of the cellulose extraction process from pineapple waste feedstocks and the modification to form cellulose nanocrystals.

inexpensive and sustainable source of biomass [45]. The cellulose extraction process for tea leaves requires repeated alkali treatments; this is likely due to the relatively high hemicellulose content relative to the cellulose content (68% vs. 16%). Therefore, the value may be in the products from hydrolysis rather than recovering the low fraction of cellulose in the native material (**Figure 6**).

4.6 Kenaf fiber

Kenaf is native to subtropical and tropical regions of Africa and Asia and holds several advantages: a short cultivation cycle, adaptability to diverse environmental conditions, and reduced requirements for pesticides and herbicides [80]. The kenaf stem has a cylindrical structure consisting of two types of fibers: the core (inner fiber) and the bark (outer fiber), which are comparable to fibers found in hardwood. Kenaf fiber is widely recognized as a valuable reinforcement material for composites in various industries [81]. Its chemical composition includes lignin (5.9–19%), hemicellulose (15–23%), and cellulose (44–63.5%) [80]. Notably, it can grow to a height of up to 3 m with a base diameter of 3–5 cm within just 3 months [81].

The extraction of cellulose from kenaf is resource-intensive [62, 81, 82]. An alkali pretreatment with 15 wt% NaOH and 1 wt% anthraquinone at 160°C is needed, followed by three bleaching steps ranging from 90 min to 3 h. Even after these pretreatments, significant milling is needed to obtain the final cellulose fiber. Therefore, assessing the cost and environmental impact of the extraction process for this feedstock is necessary (**Figure 7**).

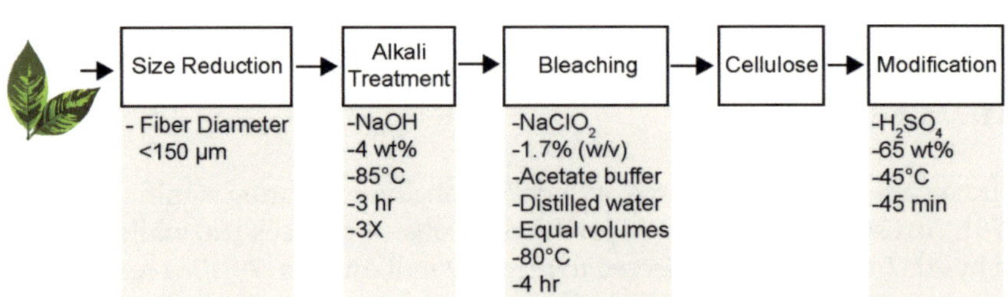

Figure 6.
Overview of the cellulose extraction process from tea leaf waste feedstocks and the modification to form cellulose nanocrystals.

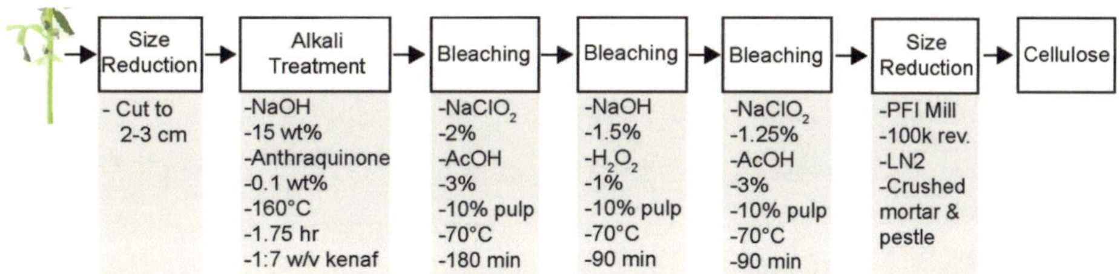

Figure 7.
Overview of the cellulose extraction process from kenaf fiber feedstocks.

4.7 Oranges

In 2024, the US production of oranges reached approximately 2.7 million tons, an increase from 2.5 million tons in the previous year [83]. Around 18% of total citrus fruit production is allocated globally for industrial purposes. However, with the growing demand for citrus fruits, the associated waste generation has also increased, contributing to environmental burdens [76, 77].

Only about 50% of the fresh orange weight is converted into juice during orange juice production, while the remaining 50% consists of peels, tissues, and polysaccharides, such as cellulose and lignin [84]. Orange peel is often discarded in landfills, exacerbating environmental concerns. Bicu and Mustata described the cellulose extraction from navel orange peels using ethylenediaminetetraacetic acid (EDTA), a process employed due to the high ash content observed when sulfite reagents are used [44]. This method increased cellulose yield and improved cellulose properties (**Figure 8**).

4.8 Tomato peels

Tomatoes are one of the most important vegetable crops globally, and they are valued for fresh produce and derivative products such as ketchup, sauce, and paste. In 2022, the United States contributed approximately 6.54% of global tomato production, amounting to 12.2 million tons [85]. However, tomato processing generates significant quantities of by-products, commonly called pomace, which account for 3–5% (w/w) of the total raw tomato weight. Tomato pomace primarily consists of 56% pulp and peels and 44% seeds on a dry basis [86].

In addition, about 31% of fresh tomatoes purchased by US households are discarded, translating to an annual economic loss of approximately $2.3 billion [87].

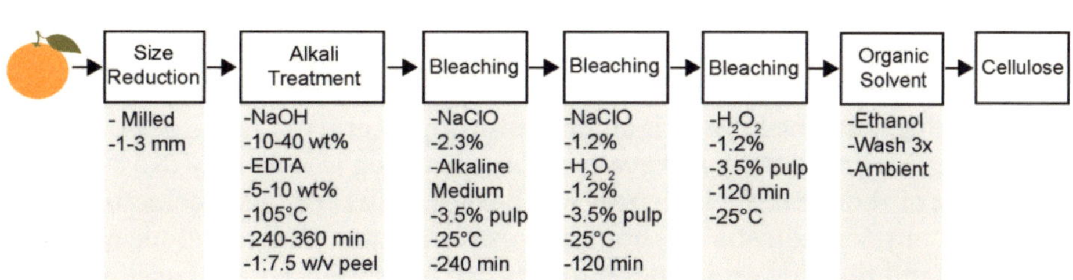

Figure 8.
Overview of the cellulose extraction process from orange feedstocks and the modification to form cellulose nanocrystals.

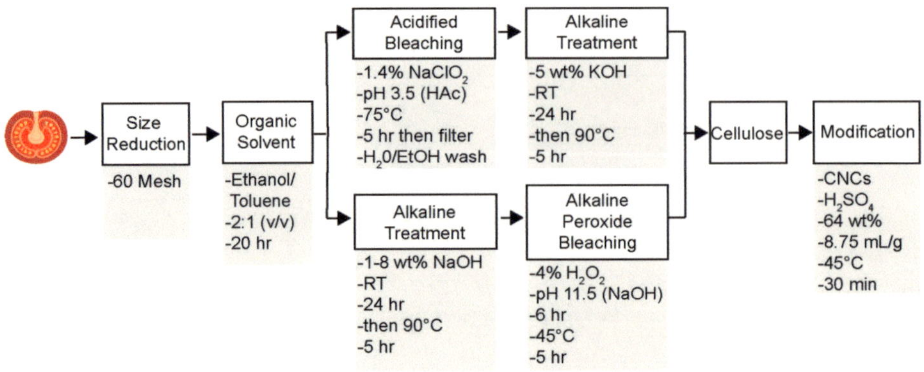

Figure 9.
Overview of the cellulose extraction process from tomato peel feedstocks and the modification to form cellulose nanocrystals.

Notably, over 50% of tomato peels, by mass, are composed of fibers, which include cellulose, hemicellulose, lignin, and pectin [59]. Despite its potential, tomato pomace currently has low-added value and is largely utilized as animal feed, compost, or disposed of in landfills, contributing to environmental challenges [86]. Recent studies have explored the production of high-value pure cellulose nanocrystals (CNCs) from tomato peels through acid hydrolysis (**Figure 9**) [59].

5. Conclusions and future direction

The increasing generation of food and agricultural waste underscores the need to find alternatives to landfills and incineration. Extracting cellulose from these waste streams offers significant environmental and economic benefits, contributing to a decrease in emissions equivalent to those produced by millions of vehicles. Despite significant research activity in this area, there is a long way to go until waste feedstocks can compete with wood resources. The following sections highlight some key areas for innovation and scientific advancement.

Separation processes are needed to aid in cellulose recovery and transition from each stage of the extraction process. One-pot reactor systems, popular in green chemistry and synthetic polymer processes, could significantly reduce cellulose loss to vacuum filtration. Additionally, because many of the feedstocks are less than 50% cellulose, separation processes that enable the recovery of other valuable small molecules or biopolymer resources could help balance the cost of the extraction process.

Due to the rapid increase in the number of experimental processes and new feedstocks being described, there is an immense opportunity to use process optimization, process simulation, and even machine learning to help improve cellulose recovery. One of the greatest challenges will be describing the feedstocks and their chemical and physical conversion, but many experimental studies include composition analysis, and the pulp and paper industry can be referenced as a baseline for unit operations. Artificial intelligence and machine learning could help identify the most promising waste feedstocks for cellulose recovery, reducing the need for arduous experimental processes that ultimately lead to low yields.

Food waste is the leading material found in landfills in the United States, accounting for 24.1% of all municipal solid waste [1]. Rapid urbanization, combined with slow progress in addressing ineffective waste management practices, contributes to the accumulation of food waste [88]. In the United States, approximately 18% of food waste is generated in restaurants, with fruits and vegetables making up 46% of the waste produced along the food supply chain [1]. Food waste is primarily lignocellulosic in nature with high cellulose and lignin content and can be transformed into high-value products using similar processes to single-source waste streams [22, 88].

Garnett et al. successfully extracted cellulose from a restaurant's mixed fruit and vegetable waste [22]. The feedstock varied based on the discarded fractions from the restaurant's menu over the course of a year. Despite the variations in the feedstock, the cellulose produced from the process had highly reproducible chemical and material properties [22]. Future work should focus on improving the yield, crystallinity, and molecular weight of the extracted cellulose, but the ability to use complex feedstocks will be transformative to the production of cellulose fibers.

As the demand for cellulose-based materials grows, addressing these real-world challenges will be critical for ensuring the sustainability and efficiency of cellulose production.

Acknowledgements

The authors acknowledge financial support from NSF Award #2316351.

Conflict of interest

The authors declare no conflict of interest.

Author details

Symone Alexander* and Mariama Alidu
Department of Chemical Engineering, Auburn University, Auburn, AL, USA

*Address all correspondence to: sla0044@auburn.edu

References

[1] Davis J. Food Waste in the United States – Ballard Brief. Ballard Brief [Internet]. 2022. Available from: https://ballardbrief.byu.edu/issue-briefs/food-waste-in-the-united-states

[2] Food and Agriculture Organization of the United Nations. Global Food Losses and Food Waste. Available from: https://www.fao.org/4/mb060e/mb060e.pdf

[3] Igini M. 10 Food Waste Statistics in America [Internet]. Earth.Org. 2022. Available from: https://earth.org/food-waste-in-america/

[4] Cooper R. Food waste in America: Facts and statistics. Rubicon Circular Economy. 2023. Available from: https://www.rubicon.com/blog/food-waste-facts/

[5] Riseh RS, Vazvani MG, Hassanisaadi M, Thakur VK. Agricultural wastes: A practical and potential source for the isolation and preparation of cellulose and application in agriculture and different industries. Industrial Crops and Products. 2024;**208**:117904 Available from: https://www.sciencedirect.com/science/article/pii/S0926669023016692

[6] Tang Z, Lin X, Yu M, Mondal AK, Wu H. Recent advances in TEMPO-oxidized cellulose nanofibers: Oxidation mechanism, characterization, properties and applications. International Journal of Biological Macromolecules. 2024;**259**:129081. Available from: https://www.sciencedirect.com/science/article/pii/S0141813023059809

[7] Ditzel FI, Prestes E, Carvalho BM, Demiate IM, Pinheiro LA. Nanocrystalline cellulose extracted from pine wood and corncob. Carbohydrate Polymers. 2017;**157**:1577-1585. Available from: https://www.sciencedirect.com/science/article/pii/S0144861716313017

[8] Srivastava D, Ahopelto J, Karttunen AJ. Thermodynamic properties of crystalline cellulose allomorphs studied with dispersion-corrected density functional methods. Molecules. 2022;**27**(19):6240. Available from: https://www.mdpi.com/1420-3049/27/19/6240

[9] George J, Sabapathi SN. Cellulose nanocrystals: Synthesis, functional properties, and applications. Nanotechnology, Science and Applications. 2015;**8**:45-54

[10] Habibi Y, Lucia LA, Rojas OJ. Cellulose nanocrystals: Chemistry, self-assembly, and applications. Chemical Reviews. 2010;**110**(6):3479-3500. Available from: https://pubs.acs.org/doi/full/10.1021/cr900339w

[11] Kurniawan TW, Sulistyarti H, Rumhayati B, Sabarudin A. Cellulose nanocrystals (CNCs) and cellulose nanofibers (CNFs) as adsorbents of heavy metal ions. Journal of Chemistry. 2023;**2023**(1):5037027. Available from: https://onlinelibrary.wiley.com/doi/full/10.1155/2023/5037027

[12] Poulose A, Parameswaranpillai J, George JJ, Gopi JA, Krishnasamy S, Dominic CDM, et al. Nanocellulose: A fundamental material for science and technology applications. Molecules. 2022;**27**(22):8032. Available from: https://www.mdpi.com/1420-3049/27/22/8032

[13] Cellulose | Definition, Uses, & Facts | Britannica. Available from: https://www.britannica.com/science/cellulose

[14] Das PK, Nag DP, Debnath S, Nayak LK. Machinery for extraction and traditional spinning of plant fibres. Indian Journal of Traditional Knowledge. 2010;**9**:386-393

[15] Lee EH, Kim DG, Hong SJ, Kim JT, Shin GH. Cellulose extraction from green algae, Ulva Ohnoi, and its application to PVA-based antibacterial composite films incorporated with zinc oxide nanoparticles. Packaging Technology and Science. 2024;**37**(6):473-485. Available from: https://onlinelibrary.wiley.com/doi/full/10.1002/pts.2805

[16] Klemm D, Heublein B, Fink HP, Bohn A. Cellulose: Fascinating biopolymer and sustainable raw material. Angewandte Chemie International Edition. 2005;**44**(22):3358-3393. Available from: https://onlinelibrary.wiley.com/doi/abs/10.1002/anie.200460587

[17] Festucci-Buselli RA et al. Structure, organization, and functions of cellulose synthase complexes in higher plants. Brazillian Journal of Plant Physiology. 2007;**19**(1):1-13

[18] Mishra RK, Sabu A, Tiwari SK. Materials chemistry and the futurist eco-friendly applications of nanocellulose: Status and prospect. Journal of Saudi Chemical Society. 2018;**22**(8):949-978. Available from: https://www.sciencedirect.com/science/article/pii/S131961031830022X

[19] Tidwell T. State of Forests and Forestry in the United States. US Department of Agriculture Forest Service. 2016. Available from: https://www.fs.usda.gov/speeches/state-forests-and-forestry-united-states-1

[20] Manzano Vela DR, Villegas Freire CN, Zabala Vizuete RF, Flores Mancheno AC. Utilization of forest residues for cellulose extraction from timber species in the high montane forest of Chimborazo, Ecuador. Polymers (Basel). 2024;**16**(19):2713. Available from: https://www.mdpi.com/2073-4360/16/19/2713

[21] Mohd Thani N, Mustapa Kamal SM, Sulaiman A, Taip FS, Omar R, Izhar S. Sugar recovery from food waste via sub-critical water treatment. Food Reviews International. 2020;**36**(3):241-257. DOI: 10.1080/87559129.2019.1636815

[22] Garnett MT, Senthil Kumar HK, Beckingham BS, Alexander SLM. Extraction of cellulose from restaurant food waste. RSC Sustainability. 2024;**2**(1):170-178. Available from: https://pubs.rsc.org/en/content/articlehtml/2024/su/d3su00192j

[23] Phiri R, Mavinkere Rangappa S, Siengchin S. Agro-waste for renewable and sustainable green production: A review. Journal of Cleaner Production. 2024;**434**:139989. Available from: https://www.sciencedirect.com/science/article/pii/S0959652623041471

[24] Matin B, Krička T, Đurović A, Grubor M, Matin A, Antonović A. Lignocellulosic composition and heating value of forest and agricultural biomass: A review. 2023. Available from: https://www.cabidigitallibrary.org/doi/pdf/10.5555/20230182679

[25] Roy P, Mohanty AK, Dick P, Misra M. A review on the challenges and choices for food waste valorization: Environmental and economic impacts. ACS Environmental Au. 2023;**3**(2):58-75. DOI: 10.1021/acsenvironau.2c00050

[26] Chopra L, Manikanika. Extraction of cellulosic fibers from the natural resources: A short review. Materials Today Proceedings. 2022;**48**:1265-1270. Available from: https://www.

sciencedirect.com/science/article/pii/S2214785321057114

[27] Kong L, Zhao Z, He Z, Yi S. Effects of steaming treatment on crystallinity and glass transition temperature of *Eucalyptuses grandis* × *E. urophylla*. Results in Physics. 2017;7:914-919. DOI: 10.1016/j.rinp.2017.02.017

[28] Bitwell C, Sen IS, Luke C, Kakoma MK. A review of modern and conventional extraction techniques and their applications for extracting phytochemicals from plants. Scientific African. 2023;**19**:e01585. Available from: https://www.sciencedirect.com/science/article/pii/S2468227623000443

[29] Wang C, Lin M, Yang Q, Fu C, Guo Z. The principle of steam explosion technology and its application in food processing by-products. Foods. 2023;**12**(17):3307. Available from: https://pmc.ncbi.nlm.nih.gov/articles/PMC10486971/

[30] Ng HM, Sin LT, Tee TT, Bee ST, Hui D, Low CY, et al. Extraction of cellulose nanocrystals from plant sources for application as reinforcing agent in polymers. Composites Part B: Engineering. 2015;**75**:176-200. Available from: https://www.sciencedirect.com/science/article/pii/S135983681500013X

[31] Mittal A, Katahira R, Himmel ME, Johnson DK. Effects of alkaline or liquid-ammonia treatment on crystalline cellulose: Changes in crystalline structure and effects on enzymatic digestibility. Biotechnology for Biofuels. 2011;**4**(1):41. DOI: 10.1186/1754-6834-4-41

[32] Romruen O, Karbowiak T, Tongdeesoontorn W, Shiekh KA, Rawdkuen S. Extraction and characterization of cellulose from agricultural by-products of Chiang Rai Province, Thailand. Polymers. 2022;**14**:1830-1843. DOI: 10.3390/polym14091830

[33] Puls J, Janzon R, Saake B. Comparative removal of hemicelluloses from paper pulps using nitren, cuen, NaOH, and KOH. Lenzinger Berichte. 2006;**86**:63-70

[34] Jonoobi M, Harun J, Shakeri A, Misra M, Oksman K. Chemical composition, crystallinity, and thermal degradation of bleached and unbleached kenaf bast (*Hibiscus cannabinus*) pulp and nanofibers. BioResources. 2009;**4**(2):626-639. Available from: https://api.semanticscholar.org/CorpusID:25906924

[35] Hart PW, Houtman C, Hirth K. Hydrogen peroxide and caustic soda: Dancing with a dragon while bleaching. Tappi Journal. 2013;**12**(7):59-65

[36] Fareez IM, Ibrahim NA, Wan Yaacob WMH, Mamat Razali NA, Jasni AH, Abdul AF. Characteristics of cellulose extracted from Josapine pineapple leaf fibre after alkali treatment followed by extensive bleaching. Cellulose. 2018;**25**(8):4407-4421. Available from: https://search.ebscohost.com/login.aspx?direct=true&AuthType=shib&db=tdh&AN=130627672&site=ehost-live&custid=s4594776

[37] Freitas PAV, Santana LG, González-Martínez C, Chiralt A. Combining subcritical water extraction and bleaching with hydrogen peroxide to obtain cellulose fibres from rice straw. Carbohydrate Polymer Technologies and Applications. 2024;7:100491

[38] Klunklin W, Hinmo S, Thipchai P, Rachtanapun P. Effect of bleaching processes on physicochemical and functional properties of cellulose and carboxymethyl cellulose from young and mature coconut coir. Polymers (Basel).

2023;**15**(16):3376. Available from: https://www.mdpi.com/2073-4360/15/16/3376

[39] Wang T, Zhao Y. Optimization of bleaching process for cellulose extraction from apple and kale pomace and evaluation of their potentials as film forming materials. Carbohydrate Polymers. 2021;**253**:117225. DOI: 10.1016/j.carbpol.2020.117225

[40] Ramos E, Calatrava S, Jiménez L. Bleaching with hydrogen peroxide. A review. Afinidad. 2008;**537**:366-373

[41] Aurelia C, Murdiati A, Supriyanto, Ningrum A. Effect of sodium hydroxide and sodium hypochlorite on the physicochemical characteristics of jack bean skin (*Canavalia ensiformis*). Pakistan Journal of Nutrition. 2019;**18**(2):193-200

[42] Aridi AS, Chin NL, Ishak NA, Yusof NM, Kadota K, Manaf YN, et al. Effect of sodium hypochlorite concentration during pretreatment on isolation of nanocrystalline cellulose from *Leucaena leucocephala* (Lam.) mature pods. BioResources. 2021;**16**(2):3137-3158. Available from: https://bioresources.cnr.ncsu.edu/resources/effect-of-sodium-hypochlorite-concentration-during-pre-treatment-on-isolation-of-nanocrystalline-cellulose-from-leucaena-leucocephala-lam-mature-pods/?utm_source=chatgpt.com

[43] Couri D, Abdel-Rahman MS, Bull RJ. Toxicological effects of chlorine dioxide, chlorite and chlorate. Environmental Health Perspectives. 1982;**46**:13-17. Available from: https://ehp.niehs.nih.gov/doi/10.1289/ehp.824613

[44] Bicu I, Mustata F. Optimization of isolation of cellulose from orange peel using sodium hydroxide and chelating agents. Carbohydrate Polymers. 2013;**98**(1):341-348

[45] Abdul Rahman NH, Chieng BW, Ibrahim NA, Abdul RN. Extraction and characterization of cellulose nanocrystals from tea leaf waste fibers. Polymers (Basel). 2017;**9**(11):588. Available from: https://www.mdpi.com/2073-4360/9/11/588

[46] Silvério HA, Flauzino Neto WP, Dantas NO, Pasquini D. Extraction and characterization of cellulose nanocrystals from corncob for application as reinforcing agent in nanocomposites. Industrial Crops and Products. 2013;**1**(44):427-436

[47] Browning BL. Methods of Wood Chemistry. Vol. 2. New York: John Wiley and Sons; 1967

[48] Costa LAS, Assis D d J, GVP G, JBA DS, Fonsêca AF, Druzian JI. Extraction and characterization of nanocellulose from corn stover. Materials Today Proceedings. 2015;**2**(1):287-294

[49] Lu S, Ma T, Hu X, Zhao J, Liao X, Song Y, et al. Facile extraction and characterization of cellulose nanocrystals from agricultural waste sugarcane straw. Journal of the Science of Food and Agriculture. 2022;**102**(1):312-321

[50] Van Soest PJ, Wine RH. Determination of lignin and cellulose in acid-detergent fiber with permanganate. Journal of AOAC International. 1968;**51**(4):780-785. DOI: 10.1093/jaoac/51.4.780

[51] Taib NAAB, Rahman MR, Bin BMK, Matin MM. Chapter 4 – Advanced techniques for characterizing cellulose. In: Rahman MR, editor. Fundamentals and Recent Advances in Nanocomposites Based on Polymers and Nanocellulose. Cambridge, United States: Elsevier; 2022. pp. 53-84. Available from: https://www.sciencedirect.com/science/article/pii/B9780323857710000014

[52] Oberlerchner JT, Rosenau T, Potthast A. Overview of methods for the direct molar mass determination of cellulose. Molecules. 2015;**20**(6):10313-10341. Available from: https://www.mdpi.com/1420-3049/20/6/10313

[53] Kraemer EO, Du EI. Molecular Weights of Celluloses. Available from: https://pubs.acs.org/sharingguidelines

[54] Viscosity of Pulp (Capillary Viscometer Method) (T230-99), TAPPI Committee of the Process and Product Quality Division; Jan 1999. Available from: https://www.document-center.com/standards/show/TAPPI-T230/history/1999%20EDITION

[55] Ju X, Bowden M, Brown EE, Zhang X. An improved X-ray diffraction method for cellulose crystallinity measurement. Carbohydrate Polymers. 2015;**123**:476-481. Available from: https://www.sciencedirect.com/science/article/pii/S0144861715000156

[56] Segal L, Creely J, Martin A, Conrad C. An empirical method for estimating the degree of crystallinity of native cellulose using the X-ray diffractometer. Textile Research Journal. 1959;**29**(10):786-794

[57] French AD, Santiago CM. Cellulose polymorphy, crystallite size, and the Segal Crystallinity Index. Cellulose. 2013;**20**(1):583-588. DOI: 10.1007/s10570-012-9833-y

[58] French AD. Idealized powder diffraction patterns for cellulose polymorphs. Cellulose. 2014;**21**:885-896

[59] Jiang F, Hsieh YL. Cellulose nanocrystal isolation from tomato peels and assembled nanofibers. Carbohydrate Polymers. 2015;**122**:60-68. Available from: https://www.sciencedirect.com/science/article/abs/pii/S0144861714012673

[60] dos Santos RM, Flauzino Neto WP, Silvério HA, Martins DF, Dantas NO, Pasquini D. Cellulose nanocrystals from pineapple leaf, a new approach for the reuse of this agro-waste. Industrial Crops and Products. 2013;**50**:707-714. Available from: https://www.sciencedirect.com/science/article/pii/S0926669013004664

[61] Lu P, Lo HY. Preparation and characterization of cellulose nanocrystals from rice straw. Carbohydrate Polymers. 2012;**87**(1):564-573

[62] Zainuddin SY, Ahmad I, Kargarzadeh H, Abdullah I, Dufresne A. Potential of using multiscale kenaf fibers as reinforcing filler in cassava starch-kenaf biocomposites. Carbohydrate Polymers. 2013;**92**(2):2299-2305

[63] D'Acierno F, Michal CA, MacLachlan MJ. Thermal stability of cellulose nanomaterials. Chemical Reviews. 2023;**123**:7295-7325. Available from: https://pubs.acs.org/doi/10.1021/acs.chemrev.2c00816

[64] Gan PG, Sam ST, Abdullah MF, bin, Omar MF. Thermal properties of nanocellulose-reinforced composites: A review. Journal of Applied Polymer Science. 2020;**137**(11):48544. Available from: https://onlinelibrary.wiley.com/doi/abs/10.1002/app.48544

[65] Md Salim R, Asik J, Sarjadi MS. Chemical functional groups of extractives, cellulose and lignin extracted from native Leucaena leucocephala bark. Wood Science and Technology. 2021;**55**(2):295-313. Available from: https://link.springer.com/article/10.1007/s00226-020-01258-2

[66] Robledo-Ortíz JR, Martín Del Campo AS, Blackaller JA,

González-López ME, Pérez Fonseca AA. Valorization of sugarcane straw for the development of sustainable biopolymer-based composites. Polymers. 2021;**13**(19):3335. Available from: https://www.mdpi.com/2073-4360/13/19/3335/htm

[67] FAOSTAT ANALYTICAL BRIEF 60 Agricultural production statistics Agricultural production statistics 2000-2021 FAOSTAT Analytical Brief 60 FAOSTAT CROPS AND LIVESTOCK PRODUCTION INTRODUCTION

[68] Frías M, Villar-Cociña E, Valencia-Morales E. Characterisation of sugar cane straw waste as pozzolanic material for construction: Calcining temperature and kinetic parameters. Waste Management. 2007;**27**(4):533-538

[69] Corn in the US – Statistics & Facts | Statista. Available from: https://www.statista.com/topics/986/corn/#topicOverview

[70] Garcia-Lara S, Serna-Saldivar SO. Chapter 1 - Corn history and culture. In: Serna-Saldivar SO, editor. Corn: Chemistry and Technology. Elsevier; 2018. pp. 1-18. Available from: https://www.sciencedirect.com/science/article/abs/pii/B9780128119716000012

[71] Zabed HM, Akter S, Yun J, Zhang G, Zhao M, Mofijur M, et al. Towards the sustainable conversion of corn stover into bioenergy and bioproducts through biochemical route: Technical, economic and strategic perspectives. Journal of Cleaner Production. 2023;**400**:136699

[72] Castorina G, Cappa C, Negrini N, Criscuoli F, Casiraghi MC, Marti A, et al. Characterization and nutritional valorization of agricultural waste corncobs from Italian maize landraces through the growth of medicinal mushrooms. Scientific Reports. 2023;**13**(1):1-16. Available from: https://www.nature.com/articles/s41598-023-48252-9

[73] Sokan-Adeaga AA, Salami SA, Bolade DO, Aledeh M, Sokan-Adeaga MA, Amubieya OE, et al. Utilization of local corn (Zea Mays) wastes for bioethanol production by separate hydrolysis and fermentation. Journal of Hazardous Materials Advances. 2024;**15**:100447

[74] Total US rice production 2023 | Statista. Available from: https://www.statista.com/statistics/190470/total-us-rice-production-from-2000/

[75] Zhang W, Xu X, Yuan Y, Wang Z. Sustainable application of rice-waste for fuels and valuable chemicals-a mini review. Frontiers in Chemistry. 2023;**11**:1225073

[76] Suri S, Singh A, Nema PK. Recent advances in valorization of citrus fruits processing waste: A way forward towards environmental sustainability. Food Science and Biotechnology. 2021;**30**(13):1601. Available from: https://pmc.ncbi.nlm.nih.gov/articles/PMC8639889/

[77] Suri S, Singh A, Nema PK. Current applications of citrus fruit processing waste: A scientific outlook. Applied Food Research. 2022;**2**(1):100050

[78] Do NHN, Ho KH, Nguyen VV, Le PK. Novel recycling of pineapple leaves into cellulose microfibers by two-step grinding of ball milling and high-speed rotor–stator homogenization. Journal of Polymer Research. 2022;**29**(6):1-9. Available from: https://link.springer.com/article/10.1007/s10965-022-03081-8

[79] Çakmak TG, Saricaoglu B, Ozkan G, Tomas M, Capanoglu E.

Valorization of tea waste: Composition, bioactivity, extraction methods, and utilization. Food Science & Nutrition. 2024;**12**(5):3112-3124. Available from: https://pubmed.ncbi.nlm.nih.gov/38726441/

[80] Saba N, Paridah MT, Jawaid M. Mechanical properties of kenaf fibre reinforced polymer composite: A review. Construction and Building Materials. 2015;**76**:87-96

[81] Ramesh P, Durga Prasad B, Narayana KL. Characterization of kenaf fiber and its composites: A review. Journal of Reinforced Plastics and Composites. 2018;**37**(11):731-737. DOI: 101177/0731684418760206. Available from: https://journals.sagepub.com/doi/10.1177/0731684418760206

[82] Asumani OML, Reid RG, Paskaramoorthy R. The effects of alkali–silane treatment on the tensile and flexural properties of short fibre non-woven kenaf reinforced polypropylene composites. Composites Part A: Applied Science and Manufacturing. 2012;**43**(9):1431-1440

[83] US orange production 2024 | Statista. Available from: https://www.statista.com/statistics/193297/us-total-orange-production-since-2000/

[84] De Medina-Salas L, Giraldi-Díaz MR, Castillo-González E, Morales-Mendoza LE. Valorization of orange peel waste using precomposting and vermicomposting processes. Sustainability. 2020;**12**(18):7626. Available from: https://www.mdpi.com/2071-1050/12/18/7626/htm

[85] Tiwari J, Rai N, Reddy YS, Singh M, Singh PM. Tomato breeding for processing in India: Current status and prospects. Vegetable Science. 2022;**49**:123-132

[86] Eslami E, Carpentieri S, Pataro G, Ferrari G. A comprehensive overview of tomato processing by-product valorization by conventional methods versus emerging technologies. Foods. 2022;**12**(1):166. Available from: https://pmc.ncbi.nlm.nih.gov/articles/PMC9818577/

[87] Buzby JC, Hyman J, Stewart H, Wells HF. The value of retail- and consumer-level fruit and vegetable losses in the United States. Journal of Consumer Affairs. 2011;**45**(3):492-515. Available from: https://onlinelibrary.wiley.com/doi/full/10.1111/j.1745-6606.2011.01214.x

[88] Ravindran R, Jaiswal AK. Exploitation of food industry waste for high-value products. Trends in Biotechnology. 2016;**34**(1):58-69

Section 2

Nanocellulose

Chapter 2

Bacterial Nanocellulose as a Green Material for Food Packaging Application

Bhargavi Pathak, Sukanya Saikia, Anurag Kashyap, Runjun Sarma, Munmi Sarma and Shahida Choudhury

Abstract

The pressing environmental challenges we face today underscore the necessity for a shift towards the development of materials that are both environmentally friendly and capable of supporting sustainable development across various sectors. Among these materials, bacterial nanocellulose (BNC) has emerged as a biomaterial with significant potential owing to its exceptional physical, chemical, and biological properties. BNC, a biopolymer composed of β-1,4 linked glucose, is produced extracellularly by some bacterial species, boasting a unique nanofibrillar structure exhibiting outstanding attributes such as high porosity, elastic modulus, crystallinity, transparency, mechanical resilience, elasticity, and water retention capacity. BNC has emerged as a versatile and sustainable biomaterial with unprecedented potential in various fields, particularly in the realm of food packaging. Besides, low-cost BNC production can be achieved using various industrial and agro wastes as substrates for bacterial culture and fermentation. BNC films align with the green chemistry principles, ensuring environmental sustainability and safety to consumers. This chapter describes the synthesis and bioengineering in harnessing BNC for creating green, active, and smart solutions to address contemporary challenges in the food packaging industry.

Keywords: bacterial nanocellulose, food, packaging, biodegradable, sustainable

1. Introduction

Food packaging plays a vital role in shaping the economy of the food industry. With the increased demand for food in the growing global population, the need for food packaging also increases. Despite its primary function of maintaining food freshness and safety, it also serves an important aspect in minimizing food waste by extending shelf life, improving transportation efficiency and reducing risk of contamination [1]. However, the modern packaging industry is a heavy consumer of plastics, causing piles of non-biodegradable municipal waste. The accumulation of plastic waste poses a global threat to environmental sustainability and raises concerns about food safety and human health. These plastics can leach harmful chemicals like bisphenol A into the stored food [2]. Moreover, they are broken down into microplastics, which enter the food chain [3]. To address these alarming issues, the transition towards safer and more sustainable food packaging alternatives is a necessity. Thus, in recent times, developing and implementing bio-based packaging solutions to mitigate the risk of plastic packaging is a major thrust area. One of the potential approaches towards sustainability in food packaging is the utilization of nanocellulose derived from plant, animal and bacterial origin. Nanocellulose can be broadly characterized into: cellulose nanofiber (CNF), cellulose nanocrystal (CNC) and bacterial nanocellulose (BNC). In contrast to CNF and CNC which are basically plant-based, BNC is produced by bacteria and is considered to be chemically pure, and does not require pre-treatment to remove lignin and hemicellulose [4]. BNC forms a stable network of nanofibers comprising of cellulose chains secreted by specific BNC-producing bacterial strains during the process of fermentation and is regarded as a green material and future of smart, active, antimicrobial and edible food packaging [5]. Besides BNC, bioplastics such as polylactic acid, derived from corn starch or sugarcane, and polyhydroxyalkanoates which are microbiologically synthesized polyesters [6], seaweed based plastics [7], mycelium based packaging that utilizes the fungal hyphal network [8] are also being explored in food packaging, as an alternative to conventional plastics used in packaging. These advancements reflect a growing shift towards eco-friendly packaging solutions aimed at minimizing plastic waste and promoting environmental sustainability.

2. Mechanical and physical properties of bacterial nanocellulose

BNC has gained popularity and interest in food packaging due to its exceptional physiochemical properties. BNC consists of ultrafine nanofibrils forming a three-dimensional web-like structure held together by strong inter and intramolecular hydrogen bonds [9]. The repeating units of glucose form the building block of BNC and create a web-like structure, making it porous in nature and promoting high surface area [10]. The ultrafine nanofibrils develop a barrier that can withstand tears and punctures and makes BNC useful in absorbing and retaining moisture up to 100 times its weight [11]. BNC is known to have high tensile strength, stiffness and flexibility, making it the ideal alternative to conventional plastic polymers. Studies have shown that BNC films exhibit high tensile strength, with values ranging from 100 to 250 MPa, and Young's modulus of 138 GPa, which is comparable to some synthetic plastics [9]. This high mechanical strength is attributed to the strong inter- and intramolecular hydrogen bonding between nanocellulose fibers [9]. Compared to plant-derived cellulose, BNC has higher crystallinity (84–89%) and purity, which contribute to its superior strength and stiffness. BNC, with its nanofibrillar network, offers a balance of strength and flexibility, making it a viable alternative [9]. BNC exhibits good mechanical resilience, increasing its load-bearing capacity and making it a mechanically robust polymer for various structural and functional applications [12]. BNC exhibits exceptional barrier properties, making it a promising material for food packaging and coating material, increasing the shelf life of perishable foods [13]. Its dense nanofibrillar structure and high crystallinity collectively hinder the diffusion of gases. BNC naturally absorbs moisture due to its hydrophilic nature, leading to higher water vapor permeability (WVP) than conventional plastics. However, this property can be modified by incorporating hydrophobic coatings such as carboxymethyl cellulose (CMC) and alkyl ketene dimer (AKD), improving the hydrophobicity of BNC [14]. Effective packaging materials must act as barriers against oxygen, moisture, and other external factors. Conventional packaging materials like polyethylene terephthalate (PET) and polyethylene (PE) provide excellent gas and moisture barrier properties, but they are non-biodegradable in nature and contribute significantly to the environmental pollution [15], whereas BNC is biodegradable and biocompatible. However, its hydrophilicity limits its moisture resistance, but recent research indicates that combining BNC with hydrophobic polymers such as poly-(butylene adipate-co-terephthalate) (PBAT) can improve its moisture barrier while retaining its biodegradable nature [16]. The native BNC has been used as a biodegradable edible packaging and can be broken down by enzymatic reactions, making it a suitable alternative for food packaging [9]. Unlike petroleum-based plastics that persist in the environment for centuries, BNC degrades into harmless byproducts such as glucose, water, and carbon dioxide [15].

3. Synthesis and production of BNC

3.1 Bacterial species used for BNC production

BNC can be biosynthesized by various bacteria, which is secreted out of the cells and acts as a survivalist trait [17]. For the bacteria, BNC acts as a nanocarrier that protects cells from harmful UV radiation. The cellulose biofilm formation also helps in moisture retention and gaseous exchange of CO_2 liberated during tricarboxylic acid cycle. Some of the notable BNC producing bacterial species are *Komagataeibacter xylinus, Komagataeibacter intermedius, Komagataeibacter saccharivorans, Komagataeibacter pasteurianus, Novacetimonas hansenii,* etc. However, a few other species of gram-negative bacteria belonging to the genera *Rhizobium*, *Agrobacterium*, *Pseudomonas*, *Azotobacter*, *Salmonella, Acaligenes*, *Achromobacter*, *Zooglea*, *Escherichia* as well as *Sarcina*, which is a gram-positive bacterium, also secretes this extracellular polymer during the fermentation process [18]. The selection of different bacterial species and their comparative analysis plays a crucial role in determining the quality of production rate and yield of the BNC [19]. The high productivity of this nanoscale cellulose production relies on a specific operon and conversion rate of glucose molecules into cellulose, indicating a high metabolic flux of carbon from glycerol, glucose and fructose [20]. During the cellulose production in *K. xylinus*, two different types of cellulose are synthesized; Cellulose I and Cellulose II, respectively. Although Cellulose II is crystalline and thermodynamically more stable than Cellulose I, it appears less frequently. In cellulose I, β-1-4 glucan chains are aligned parallel to each other, whereas in Cellulose II these chains are arranged in an antiparallel manner. The crystalline form of Cellulose I consist of two different structures: Cellulose Iα (Triclinic) and Cellulose Iβ (Monoclinic). The triclinic form is mostly produced by bacteria and algae, while the monoclinic form is more common in plants and tunicates [21]. The comparative analysis of the different strains used in BNC production possesses both advantages and disadvantages that can significantly influence their suitability in research and industrial applications. For instance the most extensively studied *K. xylinus* typically precultured in liquid Hestrin-Schramm (HS) media at 28–30°C, 5.5 pH, produces high yielding pure nanocellulose however, they are quite sensitive to environmental conditions such as pH, temperature and are solely dependent to nutrient-rich media for their optimal growth, thus increasing its rate of production [22]. *Novacetimonas hansenii,* on the other hand, under static culture has the ability to produce dense crystalline BNC films using a wide range of carbon sources that include agricultural waste and fruit-based substrates but are subjected to slower growth and lower yield as compared to *K. xylinus* [23]. *Komagataeibacter medellinensis,* which requires a low pH 3.3 for its optimal growth, particularly favors fermentation of acidic waste substrates, yet this species has limited industrial data due to its low yield [24]. Similarly, *Komagateibacter sucrofermentans* is susceptible to the risk of contamination during prolonged static cultures, despite of producing high-yielding BNC films by utilizing sucrose-rich substrates [25].

3.2 BNC production methods (static and agitated culture)

BNC synthesis is a complex enzymatic reaction process [26] with cellulose synthase as the key enzyme involved in the oxidative pentose phosphate pathway, considering glucose as the chief carbon source. The variation in the carbon sources is linked with the efficacy of cellulose production. Bacterial cellulose synthase A (BcsA) and Bacterial cellulose synthase B (BcsB) are complexes of cellulose synthase embedded in cellular membrane that polymerizes the phosphorylation of glucose molecules into glucose 6-phosphate, facilitated by glucokinase. The latter so formed is converted to glucose 1-phosphate by phosphoglucomutase, followed by production of uridine diphosphate glucose (UDP-Glc) catalyzed by UDP-pyrophosphorylase, which ultimately results in formation of cellulose chains forming bundles of well-organized and crystallized microfibrils, giving rise to a three-dimensional network of nanocellulose [27].

The production of BNC occurs in three key stages: inoculation, growth and isolation. During the inoculation stage, bacteria are introduced into a nutrient-rich growth media consisting of adequate carbon and nitrogen sources that support fermentation. The composition of growth media ensures the quality of the film produced by the bacteria. In the isolation stage, it is extracted from the growth media and processed into different forms, including sheets, films or powder, depending on the intended applications [28]. Static, agitated and bioreactor-based bacterial culturing methods are recent approaches used in BNC production [27]. In static culture, a thick leather-like hydrogel pellicle is formed at the air-liquid interface after incubating the bacterial culture in specific growth media for 7–20 days at a temperature of 25–30°C (**Figure 1**). Due to its simplicity and provision of low shear force environment it is an ideal method used in laboratories. The thickness of BNC depends on the duration of the culture process. Once the membrane is formed, it undergoes purification by being washed in a 1% sodium hydroxide solution, neutralized with acetic acid, and rinsed with deionized water until it becomes transparent. After purification, the membranes can be either dried or stored in a moist state. Static culture gives a low yield of BNC and to counter this various approaches has been made to improve the yield potential such as optimization of culture media using agro-waste which includes fruit peels, molasses, additives like glycerol, citric acid and culturing the bacterial species on trays, tanks to maximize the surface area thereby, increasing oxygen exposure. According to a recent study, it was found that by systematically varying parameters such as fermentation time, temperature and pH, a maximum cellulose yield of 8.5 g/L under static conditions can be attained at 28°C and pH 6.0 over a 9-day period [29]. Agitated cultures are used to improve the production rates of BNC by producing the cellulose in a dispersed manner within the membrane, leading to the formation of irregular pellets or suspended fibers (**Figure 1**) [30]. Bioreactor-based culture methods employ continuous BNC production using either a rotating disc (Rotary disk Bioreactor) or an airlift system with consistent oxygen transfer, in contrast to static

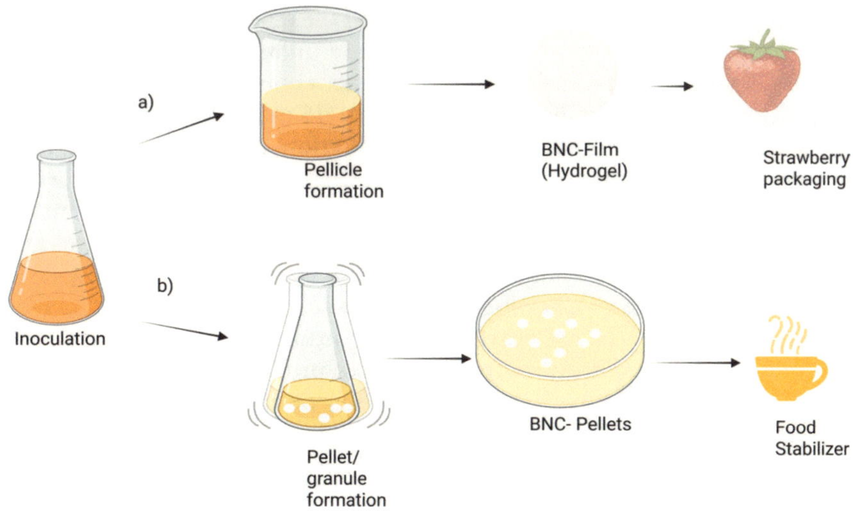

Figure 1.
Illustration showing different methods of BNC production: (a) Static culture method produces dense, crystalline, highly uniform cellulose membrane at the air/liquid interface which can be utilized in food packaging. (b) Agitated culture produces BNC aggregates/granules/pellets which can be utilized as stabilizers in various food and beverage industries [9].

culture which results in insufficient oxygenation. The membranes produced by this energy-efficient approach exhibit superior water-holding capacity and are known to have demonstrated increased yields [31].

There is ongoing research regarding the best-suited culture methods. Static culture occurs in the nutrient air-liquid interface, producing highly stable yields of BNC. In contrast, in an agitated culture, BNC synthesis begins at the particle's centre and expands outward [32]. To enhance BNC productions researchers have developed various reactor designs to maintain continuous fermentation which include: aerosol bioreactors [33], horizontal lift reactor [34], air lift reactor and spherical bubble column bioreactor [30]. The Static Intermittent Fed-Batch (SIFB) is a special type of bioreactor developed to culture the bacterial species under static conditions without agitation providing a stable undisturbed environment conductive to the formation of high-quality cellulose structure and subsequently transitions into an intermittent fed-batch mode. This strategy not only sustains the bacterial growth but also enhances BNC yields by 13.6 ± 0.32 g/L [35]. These reactors are commonly used due to their efficient oxygen transfer rates and gentle mixing, creating aerobic conditions that promote bacterial cellulose synthesis. These bioreactors operate based on gas–liquid interactions.

3.3 Substrates and media

BNC production relies on a nutrient source to sustain bacterial growth during cultivation. The most commonly used Hestrin–Schramm (HS) media consists of 2.0% glucose, 0.5% bacto-peptone, 0.5% yeast extract, 0.27% citric acid and 0.27% disodium phosphate with an initial pH of 6 [36]. Glucose, along with proteins and other nutrients, is a commonly used carbon source for energy and cellulose synthesis, typically obtained from plant-based materials like corn or sugarcane. However, due to increasing food demand and to preserve resources, glucose is not a sustainable option for long-term use. Yeast extract provides all the essential growth factors for bacterial metabolism, while peptone is the source of organic nitrogen, peptides and amino acids. Citric acid and disodium phosphate act as buffers to regulate pH during the fermentation process. Despite its effectiveness, HS media has its own limitations, as it is costly, requires substantial energy for production, and its large-scale use can have negative environmental impacts. To address these challenges, researchers have proposed more sustainable alternatives for BNC production. One of the common approaches to mitigate these issues is to replace the carbon-nitrogen sources with cheaper alternatives. These alternatives include agricultural byproducts like fruit peels, coconut water, molasses, kombucha-based cultures, using fermented tea and symbiotic culture of bacteria and yeast (SCOBY), pre-treated industrial wastewater and fruit juice waste are solutions to foster cellulose production. Utilization of these resources contributes to circular economy principles by aligning environmental consciousness with waste valorization strategies [36].

3.3.1 Valorization of Agri-waste to produce BNC film

The agro-industrial waste consists of leftover materials generated during food and agricultural production. These byproducts include potato peels, trimmings from fruits and vegetables and grain husk. Such waste material consists of significant amounts of carbohydrates, proteins, and other essential nutrients that help in the growth and production efficiency of different bacterial strains used in BNC production. Utilization of agro-waste can reduce the reliance on chemical substrates and glucose, thus lowering the production costs and promoting resource efficiency. Research has shown that certain agro-waste materials like banana peels, fruit juices, molasses and other plant-based byproducts can enhance BNC production, leading to higher yields at a lower cost [31]. These waste materials are pre-treated through various enzymatic hydrolysis, acid treatment or boiling to release simple sugars like glucose and fructose serving as alternative carbon sources. Around 100 g of pineapple and banana peel extract per liter of culture media yields 2–3.5 g/L BNC. Similarly, 20-30 g/L molasses from sugarcane can yield 4-6 g/L BNC. Recently, it was found that *K. intermedius* SB110, isolated from pineapple extract when cultured on coconut water medium enriched with sucrose and pineapple peel extract, yielded 5.563 g/L BNC in 14 days [37]. Agro-waste are inexpensive and readily available which makes it cost cost-effective and sustainable approach in BNC production. However, using these materials as substrate media can also affect the BNC quality due to their inconsistency in composition and the risk of contamination during pre-treatment stages.

4. Applications of bacterial nanocellulose (BNC) in food packaging

Bacterial nanocellulose (BNC) films have emerged as sustainable alternatives to synthetic packaging materials, offering biodegradability, mechanical strength and durability. Their potential is further modified by incorporating composite formulations to improve their protective properties (**Table 1**).

4.1 Active packaging: Antimicrobial and antioxidative properties

Active packaging solutions depend on the antimicrobial and antioxidant capabilities of composite films to help preserve food quality. Active packaging extends the

BNC film producing bacterial species used in food packaging	Additives in BNC film	Functionality observed	Type of food packaged	Ref
Komagataeibacter xylinus	Anthocyanin extracted from *Echium amoenum*	Antioxidant and pH-sensitive indicator for monitoring freshness in foods	Protein-rich food like shrimps	[42]
Gluconacetobacter xylinus	Flavonoids-Silymarin (SMN) and Zein nanoparticles	Antioxidant activity, prevents retardation of lipid oxidation	Fish products	[52]
Komagataeibacter hansenii/ *Komagataeibacter saccharivorans/* *Komagataeibacter rhaeticus*	Essential oils (EO) Thyme Cinnamon Oregeno EO's	Antimicrobial activity against pathogenic bacteria like *E. coli*, *Campylobacter*, etc.	Fresh fruits like Raspberry	[38]
Komagataeibacter xylinus	Polypyrrole/propolis extract /Zinc oxide (ZnO)	Antimicrobial activity, control of an increase in pH, increases shelf life	Meat-based products like chicken thigh meat	[53]
Komagataeibacter xylinus/ *Komagataeibacter medillinensis*	Chitosan/gelatin/probiotic bacteria	Reduces oxidative and microbial spoilage	Chicken fillets	[54]
Komagataeibacter xylinus	Nanotitanium oxide (TiO2)/copper oxide (CuO)	Phytocatalytic degradation of organic pollutants, excellent anti-fogging property, inhibition of softening, reddening and rotting of products	Fresh cut pepper/fruits and vegetables	[55]
Komagataeibacter xylinus	Cellulose nanocrystals/gellan gum/gallic acid	Antimicrobial and antioxidative activity, reduces water loss, helps in extension of shelf life	Mushroom	[56]
Komagataeibacter xylinus	Chitosan/polyvinyl alcohol (PVA)	Improves mechanical strength and flexibility, prevents invasion from mango-associated post-harvest pathogen	Mango	[57]
Komagataeibacter sucrofermentans/ *Komagataeibacter xylinus/* *Komagataeibacter hansenii/* *Komagataeibacter intermedius*	Sunflower meal protein isolates (SMPI)/crude glycerol	Reduces water loss and maintains weight of fresh fruits, antioxidant and antimicrobial activity	Strawberry	[58]
	Guar gum/polyethylene glycol (PG)/surface modified crystalline nanocellulose (NC)	Restrict growth of harmful pathogens like *E. coli*, *B. subtilis* (antimicrobial activity), enhance barrier and mechanical properties of BNC films	Grapes/fresh fruits	[59]
Gluconacetobacter xylinus	Fungal extract laccase from *Trametes versicolor*	Antioxidant and Antimicrobial activity	Fruits and vegetables	[60]

BNC film producing bacterial species used in food packaging	**Additives in BNC film**	**Functionality observed**	**Type of food packaged**	**Ref**
Komagataeibacter xylinus	e-Polylysine (e-PL)	Prolongs shelf life	Sausages/Processed food	[61]
Gluconacetobacter xylinus	Bovine lactoferrin	Promotes bactericidal effect against *Escherichia coli* and *Staphylococcus aureus*	Edible antimicrobial food packaging	[62]
Gluconacetobacter xylinus K3	Nisin	Inhibit growth of *Listeria monocytogenes (bacteria)*	Processed meat	[63]
Gluconobacter xylinus 53,582	Silver/alginate-molybdenum trioxide (Ag/Alg-MoO3) (nanoparticles)	Antimicrobial activity, improves oxygen and water vapor barrier capacity, hydrogen sulfide (H_2S) sensing ability (gas sensor)	Meat Products	[9]
Gluconacetobacter xylinus	Curcumin isolated from turmeric (Plant extract)	Photodynamic antioxidant and antibacterial activity, pH sensor, detects the presence of boric acid	Protein-based food products like meat	[64]
Gluconacetobacter xylinus	Chitosan/cellulose nanofiber/ cyclodextrin/curcumin	Reduce microbial invasion and water loss, slow down the rate of oxidation, extend shelf life	Banana, tomato, and cut apple slices	[46]
Gluconacetobacter xylinus	Starch-nanocellulose/ polyhexamethylene biguanide (PHMB)	Antibacterial activity and increases the thermal stability of the films	Grapes	[65]
Gluconacetobacter xylinus	Grafting of 3-aminopropyl trimethoxysilane onto BNC film	Antimicrobial activity	Fruits and vegetables	[1]
Komagataeibacter xylinus	Pomegranate peel extract/green tea extract/rosemary extract	Antioxidant and antimicrobial activity increases shelf life	Button Mushroom (*Agaricus bisporus*)	[66]
Komagataeibacter xylinus	Poly(sulfobetaine methacrylate) PSBMA	Optically transparent zwitterionic films with increased thermal stability, high mechanical properties, and barrier properties against UV radiation, humidity sensors	Fresh fruits like berries, leafy greens/meat, poultry, seafood/Tea bags	[67]
Komagataeibacter xylinus 4016	Lauric acid	Reduces growth of pathogenic bacteria like *Bacillus* spp. biodegradable, monitors moisture retention	Fresh produce/meat, poultry, seafood/dairy products/ice-cream packaging	[68]

BNC film producing bacterial species used in food packaging	Additives in BNC film	Functionality observed	Type of food packaged	Ref
Komagataeibacter xylinus	Polypyrrole/titanium oxide-silver (TiO2-Ag) nanocomposite	Detects plant pathogenic bacterial count, antimicrobial activity against gram-negative bacteria, and acts as a biosensor	Beverages	[69]
Gluconobacter xylinum CCM 3611	Guar gum/polyvinyl pyrrolidone (PVP)-carboxymethyl cellulose (CMC)	Hydrogel films with enhanced barrier and hydrophobic properties	Blueberries	[70]

Table 1.
Different bacterial species used in BNC film production, various modifications in films and application in food packaging.

shelf life of food products by influencing processes such as lipid/protein oxidation, microbial spoilage, and chemical deterioration. Chitosan-BNC composites exhibit strong antimicrobial properties due to their cationic charges, which enable them to interact with negatively charged microbial cell walls, ultimately causing structural damage and cell disruption [27]. Metal-based nanocomposites (e.g., silver (Ag), zinc oxide (ZnO) and titanium dioxide (TiO2)) disrupt microbial growth by producing reactive oxygen species (ROS) [9]. Physical absorption of lysozyme into the BNC membranes led to a noticeable boost in their antioxidant activity, likely due to the presence and activity of the NH2 group in the enzyme. These amino groups are known for their free radical-scavenging properties. The use of these bio-composites in the food packaging revealed that they remained stable at room temperature for several weeks with no major loss in lysozyme activity over time. Adding essential oils like rosemary and oregano, along with nanoparticles such as zinc oxide (ZnO) and silver (Ag), has been shown to effectively combat harmful bacteria like *Escherichia coli* and *Staphylococcus aureus*. Essential oil-infused BNC biofilms (e.g., oregano, thyme and citrus) provide natural antimicrobial protection [38]. BNC loaded with lactoferrin, ε-polylysine, lysozyme, or nisin actively reduces pathogens *like E. coli*, *S. aureus* and *Listeria monocytogenes*. Likewise, plant-based extracts (e.g., green tea, rosemary and pomegranate peel) enhance oxidative stability and prolong food freshness. Curcumin-BNC composites provide dual antioxidant and pH-indicating functionalities, which are beneficial for detecting spoilage. Propolis extract-enriched BNC films contain high polyphenol and flavonoid content, offering sustained antioxidant activity. Flavonoid silymarin serves as an antioxidant, effectively preventing salmon spoilage and reducing lipid oxidation, thus extending the shelf life of packaged fish [9]. Similarly, *Scrophularia striata* Boiss. extract (SE) demonstrates strong antioxidant activity, enabling the controlled release of antioxidant compounds for prolonged protection [39].

4.2 Intelligent packaging: pH and gas sensors for freshness monitoring

Intelligent packaging is a novel approach designed to monitor the quality and freshness of food products in real time. Unlike active packaging, which interacts with food to extend its shelf life, intelligent packaging primarily monitors biochemical changes within packaged food. It does this through pH indicators, gas sensors, and biosensors that provide visual or electronic feedback regarding food spoilage. During food spoilage, bacteria and enzymatic reactions produce alkaline metabolites, such as biogenic amines, ammonia, dimethylamine, and trimethylamine, which accumulate inside the package. These compounds increase the pH, signalling food deterioration [40]. BNC films have been widely used as pH-sensitive freshness indicators by incorporating colorimetric dyes. These indicators change color in response to shifts in pH, allowing consumers and retailers to assess food quality visually. Anthocyanin-based indicators such as black carrot anthocyanins were incorporated into BNC films and used as pH-sensitive indicators for monitoring fish spoilage. The indicator transitioned from red (pH 2) to gray (pH 11) [41]. *Echium amoenum* extract, rich in anthocyanins, was used to fabricate BNC-based freshness labels for shrimp packaging. The color changed from violet to yellow as pH increased, indicating microbial spoilage [42]. Embedding synthetic dye indicators such as bromophenol blue in BNC indicates freshness for mangoes. Methyl red BNC films can detect pH variations in packaged chicken, changing from red (acidic) to yellow (basic) as the spoilage advances.

Additionally, these composites were developed to detect volatile amines emitted from decomposing meat and seafood [43]. Silver nanoparticle (AgNP)-modified BNC films were created to detect H_2S gas, a major spoilage indicator for protein-rich foods. The reaction between H_2S and AgNPs resulted in colorimetric changes, signalling food deterioration [44, 45]. Additionally, introducing AgNPs into the BNC led to a remarkable increase in mechanical strength, jumping by 580% to 83.4 MPa. Incorporation of AgNPs demonstrates excellent antimicrobial activities and improved mechanical properties, making it effective and safe for potential use in packaging applications [11]. BNC/TiO_2–Ag nanocomposite films were developed as conductometric biosensors for detecting pathogenic bacterial growth. These sensors measured changes in electrical conductivity caused by microbial metabolic activity, providing a quantitative assessment of food spoilage [9]. In addition to sensing spoilage, BNC films also play a role in gas permeability, which is crucial for maintaining food quality. Oxygen permeation can accelerate food spoilage, leading to nutrient loss and degradation of food quality. The gas permeability depends on two key factors: the solubility of the gas in the film and the rate at which the gas diffuses through the film. The barrier performance in nanocellulose, including BNC has been improved by increasing the diffusion path length for gases and water vapor, effectively slowing their movement through the film. Importantly, relative humidity (RH) significantly influences the gas barrier. At a relative humidity (RH) of around 70%, strong hydrogen bonding within the nanocellulose matrix enhances cohesion and reduces gas permeability. However, at higher humidity levels, the absorption of water increases, which opens up the polymer network, allowing more gases to permeate [11]. Notably, studies have shown that at 80% RH, the oxygen transmission rate (OTR) of the film can increase up to 20-fold [46] primarily because water plastisizes the matrix, as water molecules infiltrate the structure, they weaken the intramolecular cohesion, thereby accelerating gas diffusion [47]. To address these challenges, extensive research has been conducted to explore the use of nanocellulose-based packaging materials to further enhance barrier properties [11].

4.3 BNC-composite films with additives for enhanced packaging functionality

The integration of different additives into BNC-based composite films has greatly enhanced their functionality, making them highly effective for food packaging applications (**Figure 2**). However, one of the important drawbacks of native BNC film is the lack of flexibility, which can be a major limitation for certain food packaging applications. To overcome this, BNC can be combined with polymers like polyvinyl alcohol (PVA), a non-toxic, biodegradable polymer known for its strong chemical resistance, good water absorption, effective gas barrier properties and excellent film forming abilities [48]. Furthermore, adding nanocellulose to a polyvinyl alcohol matrix significantly boosted the mechanical strength of the film by 123%, reaching 12.3 MPa compared to pure PVA [11]. These enhancements improve mechanical durability, barrier efficiency, and bioactive properties, thereby prolonging the shelf life of packaged goods while ensuring environmental sustainability [9]. The addition of biopolymers such as polyvinyl alcohol, gelatin, and pectin to BNC matrices has been shown to increase tensile strength and flexibility while preserving its lightweight nature. Studies have demonstrated that these biopolymers integrate seamlessly with BNC, forming highly stable structures that enhance the thermal resistance of the films [45]. Likewise, to enhance their ability to block moisture,

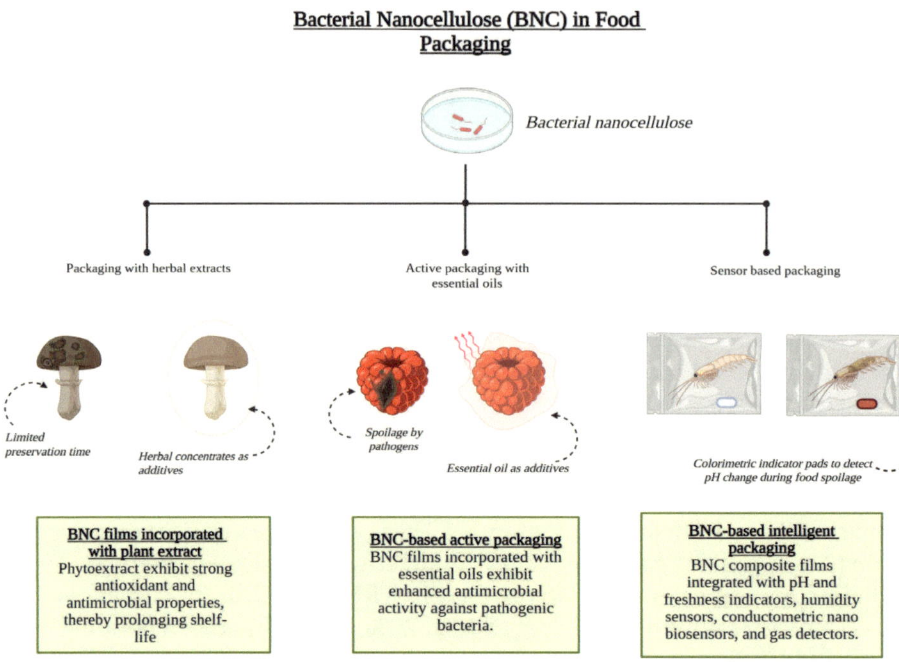

Figure 2.
Illustration showing application of BNC in active and intelligent packaging of food products for enhanced shelf life, reduced pathogen damage and real-time monitoring of the quality of the food packed.

oxygen, and other environmental elements, BNC-based composite films often need additional reinforcement. Adding materials like nanofibrillated cellulose (NFC), carboxymethylcellulose (CMC), and poly (butylene adipate-co-terephthalate) (PBAT) has proven effective in lowering water vapor transmission rates (WVTR) and oxygen permeability. Two primary strategies have been explored for the functionalization of bacterial cellulose: *in situ* and *ex situ* modification methods [17]. The more commonly employed technique is the *in situ* approach, where the reinforcing agents like sodium alginate, CMC, polyvinyl alcohol, gelatin, agar, pectin or starch are directly incorporated into the culture medium during the early stages of BNC biosynthesis [9]. This method is advantageous due to its simplicity and ability to integrate additives into the forming BNC network, leading to stable composites with tailored properties. However, the drawback of this technique is poor solubility of some additives in the culture medium and the effects of certain actives on bacterial growth [17]. Alternatively, in *ex situ* technique, the bioactive agents are impregnated in the dried or hydrated BNC matrix. Nevertheless, it is constrained by its incompatibility with hydrophobic compounds and the limited penetration of the larger particles, as only submicron and nano-sized agents can diffuse into the porous BNC network [49]. Despite these limitations, both *in situ* and *ex situ* functionalization techniques have significantly enhanced the physiochemical properties of the native BNC, expanding their use in the advanced food packaging solution [9]. These improvements make BNC films a strong alternative to conventional plastic packaging, offering both durability and sustainability [46]. These additives not only strengthen the film's mechanical properties but also promote environmentally friendly decomposition, making them a sustainable alternative to traditional plastics with a lower ecological footprint [15].

5. Safety and regulatory consideration

BNC has emerged as a sustainable biomaterial with applications in food packaging, biomedical engineering, pharmaceuticals and cosmetics. BNC is widely recognized for its non-toxicity, biocompatibility, and biodegradability, tailoring it for direct food contact and biomedical applications. The US Food and Drug Administration (FDA) has granted Generally Recognized as Safe (GRAS) status to BNC for food-related applications [47]. BNC, due to its inert nature and high purity, does not release harmful chemicals or interact negatively with packaged food and has been historically used in food products such as nata de coco, confirming its safety for human consumption [48]. Studies have confirmed that BNC membranes do not trigger inflammatory responses, making them suitable for wound dressings, drug delivery systems, and tissue engineering scaffolds [50]. However, traces of bacterial endotoxins such as (1,3)-β-D-glucan from the fermentation process can pose immunogenic risks, requiring rigorous purification protocols before biomedical applications [49]. In the United States, various agencies are responsible for regulating nanocellulose in different sectors. The Institute for Food and Agricultural Standards (IFAS), the Food and Drug Administration (FDA), and the US Environmental Protection Agency (USEPA) oversee safety assessments, production guidelines and commercial applications of nanocellulose-based materials. Similarly, in the European Union, several regulatory bodies such as Registration, Evaluation, Authorisation and Restriction of Chemicals (REACH), European Region on Classification, Labeling and Packaging (CLP) of Chemicals, Scientific Committee on Emerging and Newly Identified Health Risks (SCENIHR), establish standards for the safe use nanocellulose, particularly in biomedical and environmental applications. Other international organizations such as the Scientific Committee on Consumer Safety (SCCS), the International Cooperation on Cosmetic Regulation (ICCR), national authorities such as the UK Soil Association, the Biological Farmers of Australia, and the Canada General Standards Board ensure the safe incorporation of nanocellulose in the industrial, biomedical and environmental applications [51].

6. Conclusion and future prospects

BNC is emerging as a promising alternative to conventional food packaging materials due to their outstanding mechanical, physical, and barrier properties. BNC can also be blended with other biopolymers to enhance its characteristics and broaden its applications. Many research efforts have focused on minimizing production costs by improving the composition of culture media with more affordable alternatives, such as residual materials like agro-waste. A major challenge in commercializing BNC is scaling up its production. Currently, the fermentation process yields only small amounts of BNC, making large-scale production challenging. This limitation poses a significant barrier to the commercialization of BNC, and overcoming it will require future research and development. However, the development of new technologies like bioreactors enhances BNC's accessibility and market competitiveness. Recently, a static intermittent fed-batch (SIFB) method utilizing low-cost media, including agro-based products has been introduced. This approach has demonstrated a high substrate conversion rate to BNC, nearly doubling the yield compared to conventional static methods. Significant variations in the physiochemical properties of BNC were observed due to differences in bacterial species, to address this issue, proper cataloguing is essential. Innovations in bioengineering BNC open the avenues to impose newer functionalities using nanoparticles, anthocyanins, chitosan, fatty acid, plant extract and enzymes. BNC films align with the green chemistry principles, ensuring sustainability and safety for human health, offering endless possibilities in diverse areas.

Author details

Bhargavi Pathak[1], Sukanya Saikia[1], Anurag Kashyap[1*], Runjun Sarma[2], Munmi Sarma[3] and Shahida Choudhury[4]

1 Department of Plant Pathology, Assam Agricultural University, Jorhat, Assam, India

2 Department of Physics, Haflong Government College, Dima Hasao, Assam, India

3 Department of Chemical Sciences, Tezpur University, Tezpur, Assam, India

4 College of Horticulture and Farming System Research, Assam Agricultural University, Nalbari, Assam, India

*Address all correspondence to: anurag.kashyap@aau.ac.in

References

[1] Acharyya P, Sarma M, Kashyap A. Recent advances in synthesis and bioengineering of bacterial nanocellulose composite films for green, active and intelligent food packaging. Cellulose. 2024;**31**:7163-7187

[2] Almeida S, Raposo A, Almeida-Gonazalez M, Carrascosa C. Bisphenol a: Food exposure and impact on human health. Comprehensive reviews on Food Science and Food Safety. 2018;**17**:1503-1517

[3] Ncube L, Ude A, Ogunmuyiwa E, Zulkifli R, Beas I. Environmental impact of food packaging materials: A review of contemporary development from conventional plastics to Polylactic acid based materials. Materials. 2020;**21**:4994

[4] Norrahim M, Nurazzi N, Jenol M, Abdillah M, Farid A, Janudin N, et al. Emerging development of nanocellulose as an antimicrobial material: An overview. Materials Advances. 2021;**2**:3538-3551

[5] Urbina L, Corceura M, Gabilondo N, Eceiza A, Retegi A. A review of bacterial cellulose: Sustainable production from agricultural waste and applications in various fields. Spring. 2021;**28**:8229-8253

[6] Naser A, Deiab I, Darras B. Poly(lactic acid) (PLA) and polyhydroxyalkanoates (PHAs), green alternatives to petroleum-based plastics: A review. RSC Reviews. 2021;**11**:17151-17196

[7] Hamed I, Jakobsen A, Lerfall J. Sustainable edible packaging systems based on active compounds from food processing byproducts: A review. Comprehensive Reviews in Food Science and Food Safety. 2021;**21**:196-226

[8] Abhijith R, Ashok A, Rajeesh C. Sustainable packaging applications from mycelium to substitute polystyrene: A review. Materials Today. 2018;**5**(1):2139-2145

[9] Ludwicka K, Kaczmarek M, Białkowska A. Bacterial Nanocellulose—A biobased polymer for active and intelligent food packaging applications: Recent advances and developments. Polymers. 2020;**12**(10):2209

[10] Moniri M, Moghaddam A, Azizi S, Rahim R, Ariff A, Saad W, et al. Production and status of bacterial cellulose in biomedical engineering. Nanomaterials. 2017;**7**(9):257

[11] Ahankari S, Subhedar A, Bhadauria S, Dufresne A. Nanocellulose in food packaging: A review. Carbohydrate Polymers. 2021;**255**:117479

[12] Haimer E, Wendland M, Schlufter K, Frankenfeld K, Miethe P, Potthast A, et al. Loading of bacterial cellulose aerogels with bioactive compounds by Antisolvent precipitation with supercritical carbon dioxide. Macromolecular Symposia. 2010;**294**(2):64-74

[13] Valsalan A, Sivaranjana P. Review article: Development of biodegradable films using nanocellulose for food packaging application. Analytical Methods in Environmental Chemistry Journal. 2022;**5**(4):20-42

[14] Nadeem H, Dehghani M, Miri S, Pazirofteh M, Garnier G, Batchelor W. Highly hydrophobic and moisture barrier nanocellulose based films produced via spray deposition. Cellulose. 2023;**30**(8):5157-5170

[15] Hussain M, Mishra S, Agarwal Y, Rathore D, Chokshi N. A comparative review of biodegradable and conventional plastic packaging. Springer Nature. 2024;**245**:126

[16] Kaczmarek M, Krzepkowska M, Ludwicka K. Comparative analysis of bacterial cellulose membranes synthesized by chosen Komagataeibacter strains and their application potential. International Journal of Molecular Sciences. 2022;**23**(6):3391

[17] Cacicedo M, Cristina Castro M, Servetas I, Bosnea L, Boura K, Tsafrakidou P, et al. Progress in bacterial cellulose matrices for biotechnological applications. Bioresource Technology. 2016;**213**:172-180

[18] Ross P, Mayer R, Benziman M. Cellulose biosynthesis and function in bacteria. Microbiology and Molecular Biology Reviews. 1991;**5**:35-58

[19] Samyn P, Meftahi A, Geravand S, Heravi M, Najarzadeh H, Sabery M, et al. Opportunities for bacterial nanocellulose in biomedical applications: Review on biosynthesis, modification and challenges. International Journal of Biological Macromolecules. 2023;**231**:123316

[20] Wang J, Tavakoli J, Tang Y. Bacterial cellulose production, properties and applications with different culture methods - a review. Carbohydrate Polymers. 2019;**219**:63-76

[21] Yu X, Atalla R. Production of cellulose II by Acetobacter xylinum in the presence of 2,6-dichlorobenzonitrile. International Journal of Biological Macromolecules. 1996;**19**(2):145-146

[22] Gao H, Sun Q, Han Z, Li J, Liao B, Hu L, et al. Comparison of bacterial nanocellulose produced by different strains under static and agitated culture conditions. Carbohydrate Polymers. 2020;**227**:115323

[23] Neelima S, Sreejith S, Shahjahan S, Raj A, Vidya L, Aparna V, et al. Highly crystalline bacterial cellulose production by Novacetimonas hansenii strain isolated from rotten fruit. Materials Letters. 2023;**333**:133622

[24] Hernandez-Arriaga A-M, Cerro C, Urbina L, Eceiza A, Corcuera M, Retegi A, et al. Genome sequence and characterization of the bcs clusters for the production of nanocellulose from the low pH resistant strain Komagataeibacter medellinensis ID13488. Microbial Biotechnology. 2019;**4**:620-632

[25] Adamopoulou V, Bekatorou A, Brinias V, Michalopoulou P, Dimopoulos C, Zafeiropoulos J, et al. Optimization of bacterial cellulose production by Komagataeibacter sucrofermentans in synthetic media and agrifood side streams supplemented with organic acids and vitamins. Bioresource Technology. 2024;**398**:130511

[26] Gromet Z, Schramm M, Hestrin S. Synthesis of cellulose by Acetobacter xylinum. 4. Enzyme systems present in a crude extract of glucose-grown cells. Biochemical Journal. 1957;**67**(4):679-689

[27] Bajra F. Bacterial nanocellulose production and biomedical applications. The Journal of Biomedical Research. 2021;**35**(4):310-317

[28] Ullah M, Alabbosh K, Fatima A, Islam S, Manan S, Ul-Islam M, et al. Advanced biotechnological applications of bacterial nanocellulose-based biopolymer nanohybrids: A review. Advanced Industrial and Engineering Polymer Research. 2024;**7**(1):100-121

[29] Chen J, Hong F, Zheng H, Zheng L, Du B. Using static culture method to

increase the production of Acetobacter Xylinum bacterial cellulose. Journal of Natural Fibers. 2024;**21**:2288286

[30] Skočaj M. Bacterial nanocellulose in papermaking. Cellulose. 2019;**26**: 6477-6488

[31] Lin S, Calvar I, Catchmark J, Liu J, Demirci A, Cheng K. Biosynthesis, production and applications of bacterial cellulose. Cellulose. 2013;**20**:2191-2219

[32] Reshmy R, Philip E, Thomas A, Madhavan R, Sindhu P, Binod S, et al. Bacterial nanocellulose: Engineering, production, and applications. Bioengineered. 2021;**12**(2):11463-11483

[33] Hornung M, Ludwig M, Schmauder H. Optimizing the production of bacterial cellulose in surface culture: A novel aerosol bioreactor working on a fed batch principle (part 3). Engineering in Life Science. 2007;**7**:35-41

[34] Kralisch D, Hessler N, Klemm D, Erdmann R, Schmidt W. White biotechnology for cellulose manufacturing—The HoLiR concept. Biotechnology and Bioengineering. 2009;**107**:740-747

[35] Sharma C, Bharadwaj N, Pathak P, Dey P, Gautam S, Kumar S, et al. Bacterial nanocellulose by static, static intermittent fed-batch and rotary disc bioreactor-based fermentation routes using economical black tea broth medium: A comparative account. International Journal of Biological Macromolecules. 2024;**277**:134228

[36] Schramm M, Hestrin S. Factors affecting production of cellulose at the air/liquid interface of a culture of Acetobacter xylinum. Microbiology. 1954;**11**(1):123-129

[37] Perdani C, Nurika I, Gunawan S. Characterization of bacterial Nanocellulose produced by tropical fruit-derived bacteria in coconut water medium supplemented with pineapple Peel extract. Chemistry Select. 2024;**9**(35):e202402954

[38] Casalini S, Baschetti M, Cappelletii M, Guerreiro A, Gago C, Antunes M. Antimicrobial activity of different Nanocellulose films embedded with thyme, cinnamon and oregano essential oils for active packaging application on raspberries. Frontiers in Sustainable Food Systems. 2013;**7**:11909-11979

[39] Sukhtezari S, Almasi H, Pirsa S, Zandi M, Pirouzifard M. Development of bacterial cellulose based slow-release active films by incorporation of Scrophularia striata Boiss. Extract. Carbohydrate Polymers. 2017;**156**:340-350

[40] Balbinot-Alfaro E, Craveiro D, Lima K, Costa H, Lopes D, Prentice C. Intelligent packaging with pH indicator potential. Food Engineering Reviews. 2019;**11**:235-244

[41] Moradi M, Tajik H, Almasi H, Forough M, Ezati P. A novel pH-sensing indicator based on bacterial cellulose nanofibers and black carrot anthocyanins for monitoring fish freshness. Carbohydrate Polymers. 2019;**222**:115030

[42] Mohammadalinejhad S, Almasi H, Moradi M. Immobilization of Echium amoenum anthocyanins into bacterial cellulose film: A novel colorimetric pH indicator for freshness/spoilage monitoring of shrimp. Food Control. 2020;**113**:107169

[43] Kaswandi B, Larasati J, Abdullah A, Heng L. Real-time monitoring of shrimp spoilage using on-package sticker sensor based on natural dye of curcumin. Food Anlytical Methods. 2012;**5**(1):881-889

[44] Sukhavattanakul P, Manuspriya H. Fabrication of hybrid thin film based on bacterial cellulose nanocrystals and metal nanoparticles with hydrogen sulfide gas sensor ability. Carbohydrate Polymers. 2020;**230**:115566

[45] Wahid F, Wang F, Xie Y-Y, Chu L-Q, Jia S-R, Duan Y-X, et al. Reusable ternary PVA films containing bacterial cellulose fibers and ε-polylysine with improved mechanical and antibacterial properties. Colloids and Surfaces B: Biointerfaces. 2019;**183**:110486

[46] Zhou Y, Liu R, Zhou C, Gao Z, Gu Y, Chen S, et al. Dynamically crosslinked chitosan/cellulose nanofiber-based films integrated with γ-cyclodextrin/curcumin inclusion complex as multifunctional packaging materials for perishable fruit. Food Hydrocolloids. 2023;**144**:108996

[47] Dourado F, Gama M, Rodrigues A. A review on the toxicology and dietetic role of bacterial cellulose. Toxicology Reports. 2017;**4**:543-553

[48] Azeredo H, Barud H, Farinas C, Vasconcellos V, Claro A. Bacterial cellulose as a raw material for food and food packaging applications. Frontiers in Sustainable Food Systems. 2019;**3**:7

[49] Liu J, Bacher M, Rosenau T, Willfor S, Mihranyan A. Potentially immunogenic contaminants in wood-based and bacterial Nanocellulose: Assessment of endotoxin and (1,3)-β-d-glucan levels. Biomacromolecules. 2018;**19**(1):150-157

[50] Abeer M, Mohd Amin M, Martin C. A review of bacterial cellulose-based drug delivery systems: Their biochemistry, current approaches and future prospects. Journal of Pharmacy and Pharmacology. 2014;**66**(8):1047-1061

[51] Jeevanandam J, Barhoum A, Chan Y, Dufresne A, Danquah M. Review on nanoparticles and nanostructured materials: History, sources, toxicity and regulations. Beilstein Journal of Nanotechnology. 2018;**9**:1050-1074

[52] Tsai Y-H, Yang Y-N, Ho Y-C, Tsai M-L, Mi F-L. Drug release and antioxidant/antibacterial activities of silymarin-zein nanoparticle/bacterial cellulose nanofiber composite films. Carbohydrate Polymers. 2018;**180**:286-296

[53] Pirsa S, Shamusi T. Intelligent and active packaging of chicken thigh meat by conducting nano structure cellulose-polypyrrole-ZnO film. Materials Science and Engineering. 2019;**102**:798-809

[54] Salimiraad S, Safaeian S, Basti A, Khanjari A, Nadoushan R. Characterization of novel probiotic nanocomposite films based on nano chitosan/ nano cellulose/ gelatin for the preservation of fresh chicken fillets. LWT. 2022;**162**:113429

[55] Chen F, Chang X, Xu H, Fu X, Ding S, Wang R. Gellan gum-based functional films integrated with bacterial cellulose and nano-TiO2/CuO improve the shelf life of fresh-cut pepper. Food Packaging and Shelf Life. 2023;**38**:101103

[56] Criado P, Fraschini C, Becher D, Pereira F, Salmieri S, Lacroix M. Modified Cellulose Nanocrystals (CNCs) Loaded in Gellan Gum Matrix Enhance the Preservation of Agaricus Bisporus Mushrooms. Food Processing and Preservation. Journal of Food Processing and Preservation. 2020;**44**(11):e14856

[57] Dey D, Dharini V, Selvam S, Sadiku E, Kumar M, Jayaramudu J, et al. Physical, antifungal, and biodegradable properties of cellulose nanocrystals

and chitosan nanoparticles for food packaging application. Materials Today. 2021;**38**(2):860-869

[58] Efthymiou M, Tsouku E, Papagiannopoulos A, Athanasoulia I, Georgidou M, Pispas S, et al. Development of biodegradable films using sunflower protein isolates and bacterial nanocellulose as innovative food packaging materials for fresh fruit preservation. Scientific Reports. 2022;**12**:6935

[59] Garg R, Rana H, Singh N, Goswami S. Guargum/nanocellulose based novel crosslinked antimicrobial film with enhanced barrier and mechanical properties for food packaging. Journal of Environmental Chemical Engineering. 2023;**II**(2):109254

[60] Chen L, Zou M, Hong F. Evaluation of fungal laccase immobilized on natural nanostructured bacterial cellulose. Frontiers in Microbiology. 2015;**6**:1245

[61] Criado P, Hossain F, Salmieri S, Lacroix M. Nanocellulose in food Packaging. In: Composites Materials for Food Packaging. Insight to Modern Food Science. Hoboken, NJ: Wiley; 2018. pp. 297-329

[62] Padrao J, Goncalves S, Silva J, Lanceros-Mendez S, Pinheiro A, Vicente A, et al. Bacterial cellulose-lactoferrin as an antimicrobial edible packaging. Food Hydrocolloids. 2016;**58**:126-140

[63] Nguyen V, Gidley M, Dykes G. Potential of a nisin-containing bacterial cellulose film to inhibit listeria monocytogenes on processed meats. Food Microbiology. 2008;**25**(3):471-478

[64] Yang Y-N, Lu K-Y, Wang P, Ho Y-C, Tsai M, Mi F-L. Development of bacterial cellulose/chitin multi-nanofibers based smart films containing natural active microspheres and nanoparticles formed in situ. Carbohydrate Polymers. 2020;**228**:115370

[65] Yuan Y, Chen H. Preparation and characterization of a biodegradable starch-based antibacterial film containing nanocellulose and polyhexamethylene biguanide. Food Packaging and Shelf Life. 2021;**30**:100718

[66] Moradian S, Almasi H, Moini S. Development of bacterial cellulose-based active membranes containing herbal extracts for shelf life extension of button mushrooms (Agaricus bisporus). Journal of Food Processing and Preservation. 2017;**42**(9):e13537

[67] Vilela C, Moreirinha C, Domingues E, Figueiredo F, Almeida A, Freire S. Antimicrobial and conductive Nanocellulose-based films for active and intelligent food packaging. Nanomaterials. 2019;**9**(7):980

[68] Zahan K-A, Azizul N, Mustapha M, Tong W, Rahman M, Sahuri I. Application of bacterial cellulose film as a biodegradable and antimicrobial packaging material. Materials Today. 2020;**31**:83-88

[69] Ghasemi S, Bari M, Pirsa S, Amiri S. Use of bacterial cellulose film modified by polypyrrole/TiO2-Ag nanocomposite for detecting and measuring the growth of pathogenic bacteria. Carbohydrate Polymers. 2020;**232**:115801

[70] Bandyopadhyay S, Saha N, Brodnjak U-V, Saha P. Bacterial cellulose and guar gum based modified PVP-CMC hydrogel films: Characterized for packaging fresh berries. Food Packaging and Shelf Life. 2019;**22**:100402

Chapter 3

Nanocellulose, a Promising Raw Material: Improving the Scalability of TEMPO-Mediated Oxidation Process

Jose Luis Sanchez-Salvador, Hongyu Xu, Ana Balea, Elena Fuente, M. Concepcion Monte, Angeles Blanco and Carlos Negro

Abstract

Nanocellulose is a promising raw material due to its distinctive properties, including renewable origin, biodegradability, lightweight, and high mechanical strength. It has a very high potential to enhance products in a broad spectrum of applications. However, the production of highly fibrillated cellulose nanofibrils (CNFs) remains costly due to the high energy and chemical consumption. TEMPO-mediated oxidation (TMO) is the most widely accepted pretreatment for CNF production due to its high efficiency and selectivity. However, challenges associated with scaling up this process are limiting their implementation, as high catalysts and oxidant doses, extended reaction times, and large reaction volumes. Several strategies have been developed with the aim of enhancing the CNF production and optimizing the overall process. These strategies include real-time monitoring of the reaction parameters, optimizing pulp concentration, reusing the reaction medium, and using different reactor configurations such as kneaders and twin-screw extruders. These advancements are reviewed to show the significant and critical progress carried out in the last decade toward achieving more efficient and sustainable nanocellulose production.

Keywords: cellulose nanofibrils, nanocellulose, TEMPO-mediated oxidation, twin-screw extruder, kneading, upscaling

1. Introduction

In the 1980s of the last century, nanocellulose (NC), or cellulose nanomaterials (CNMs), emerged as a new family of functional materials [1]. However, significant interest in these materials did not develop until the twenty-first century, when research on NC grew exponentially due to the need for sustainable biomaterials. NC are characterized by their biodegradability, lightweight and compatibility with other materials. Furthermore, they present high strength, high surface area, and functionalization possibilities [2]. Cellulose, the primary raw material for NC, can be obtained not only from wood and plants but also from agricultural products and residues, tunicates, or algae. Furthermore, cellulose is also produced by some microorganisms such as bacteria and fungi. Bacteria cellulose is structurally similar to plant-derived cellulose but has distinct properties such as higher purity and better water retention [3, 4].

The exceptional properties of NC allow its application across diverse fields, such as the paper industry [5], pharmaceuticals and biomedical applications [6], the food industry [7], energy storage [8], the cement industry [9], or in environmental applications [10, 11]. Each application requires a tailored study to identify the most suitable CNMs, based on the specific properties needed, following the approach of fit-for-use.

Two main characteristics define CNMs: (i) being predominantly composed of cellulose, although other compounds such as lignin or hemicellulose may also be present and (ii) having at least one external dimension in the nanoscale, under 100 nm, or an internal structure or surface structure in the nanoscale (nanostructured materials) [12].

NC is mainly produced through defibrillation of cellulose fibrils. The cellulose chains, linked by hydrogen bonds, form elementary cellulose fibrils with diameters in the nanoscale and with lengths of several micrometers. Within each elementary fibril, regions of ordered cellulose chains (crystalline) provide high strength, while disordered (amorphous) regions contribute to elasticity. These elementary fibrils are subsequently arranged into microfibers or fibers that integrate into the cell wall structure.

Historically, the terminology used in the definition of CNMs has been diverse, with many analogous terms. The terminology has been standardized by the ISO standard ISO/TS 20477:2017, defining terms and definitions for different types of CNMs, categorizing NC into two main groups based on their structure [13]. The first group, cellulose nano-objects, consists of materials containing at least one elementary fibril and an external dimension in the nanoscale. This group is subdivided into cellulose nanocrystals (CNCs), containing predominantly crystalline and paracrystalline regions with aspect ratios (AR) between 5 and 50, and CNFs, which contain both crystalline and amorphous regions with aspect ratios greater than 10. The second group, cellulose nanostructured materials, comprises aggregated bundles of cellulose with nanoscale internal or/and surface structures. This group includes cellulose microcrystals (CMCs) and cellulose microfibrils (CMFs).

Among the different CNMs, cellulose nano-objects are the most studied by presenting the separation of the elementary fibers after defibrillation. However, since the treatments performed are not always efficient enough to obtain the nano-objects (CNFs and CNCs), CMFs and CMCs could coexist in the final product.

This chapter focuses on the production of CNFs, which are defined as cellulose nano-objects with crystalline, paracrystalline, and amorphous regions with two similar external dimensions in the nanoscale, approximately from 3 to 100 nm, and the third dimension significantly larger with a cross-section up to 100 μm [13].

The most common treatment for producing CNFs involves a mechanical defibrillation of cellulose fibers, deconstructing them through high-shear forces in equipments such as homogenizers, refiners, microfluidizers, grinders, mills, or extruders [14]. Beside new combinations of treatments have been developed in recent years, where the main treatment does not rely on mechanical forces, such as electrospinning in which a high voltage is applied to a cellulose solution, which forces it to form fibers that are collected on a surface [15].

Mechanical treatments require multiple passes to completely break down the fibers into elementary fibrils, resulting in a high demand for energy and time. Several pretreatments based on chemical and enzymatic processes have been developed for enhancing fiber fibrillation and reducing energy consumption in mechanical defibrillation while enhancing process efficiency [16]. Therefore, depending on the effectiveness of the treatments, a wide range of products can be obtained with different properties and homogeneity degree [17]. Low-intensive treatments produce yield heterogeneous samples containing CNFs mixed with larger fibers in the range of CMFs. On the contrary, highly intensive treatments produce a high yield of homogeneous CNFs, as in the case of TEMPO-mediated oxidation (TMO) followed by several passes of homogenization at high pressure (HPH), which is by far the most commonly used treatment combination.

2. TEMPO-mediated oxidation

Chemical pretreatments are key for disrupting the complex matrix of the plant cell wall, partially digesting structural components, and modifying cellulose surface [4]. Among them, TMO is particularly effective, as it produces the carboxylation of cellulose, enabling electrostatic repulsion between the oxidized C6 groups of cellulose and facilitating the separation of fibrils during the subsequent mechanical treatment [18].

The oxidation of cellulose catalyzed by 2,2,6,6-Tetramethylpiperidine-1-oxyl (TEMPO) was first reported by Isogai and Kato in 1998 [19], using NaBr as a co-catalyst and NaClO as oxidant. TEMPO, a *N-oxoammonium* salt, acts as a catalyst in the oxidation of primary alcohols to aldehydes and subsequently to carboxyl groups. TEMPO acts selectively on the C6 hydroxyl groups of anhydroglucose units of cellulose (**Figure 1**). The most commonly used oxidation conditions were established by Saito et al. [21], who employed a stirring reactor with a 1% concentration of cellulose pulp, 0.1 mmol/g pulp of TEMPO, and 1 mmol/g of NaBr [21]. Since then, various NaClO doses have been tested to optimize the oxidization degree of the cellulose fibers [22, 23].

As shown in **Figure 1**, the reaction begins with the addition of the oxidant to the reactor containing the pulp and the catalysts. The pH is usually maintained around 10 by adding NaOH until the pH stabilizes when all NaClO is consumed [24]. The oxidized pulp (OP) undergoes a cleaning stage, which includes vacuum filtration through a nylon mesh, followed by several washing steps with distilled water and filtering until a neutral pH is reached. To prevent bacterial growth, a few drops of glutaraldehyde solution are added as bactericide to the OPs at 20 wt.% solid content. The OPs are then stored at 4°C until characterization and CNF production, which take place after the main mechanical treatment.

TMO pretreatment has several advantages, but it also has two main drawbacks. The first one is the partial dissolution of cellulose in the reaction medium, reducing the polymerization degree [25]. This reaction is beneficial to facilitate subsequent mechanical treatment as long as the level of β-1,4-glycosidic bond breakage is moderate. Nevertheless, a large amount of oxidant may lead to excessive chain breakage, causing the dissolution of an abundant amount of cellulose into the reaction medium and decreasing the yield. Some alternatives to avoid pulp depolymerization have been developed based on using other oxidation systems, such as TEMPO/NaClO/$NaClO_2$ in aqueous-organic solvents. However, the amount of carboxyl groups obtained is fewer than that produced under traditional conditions, which makes it difficult for the

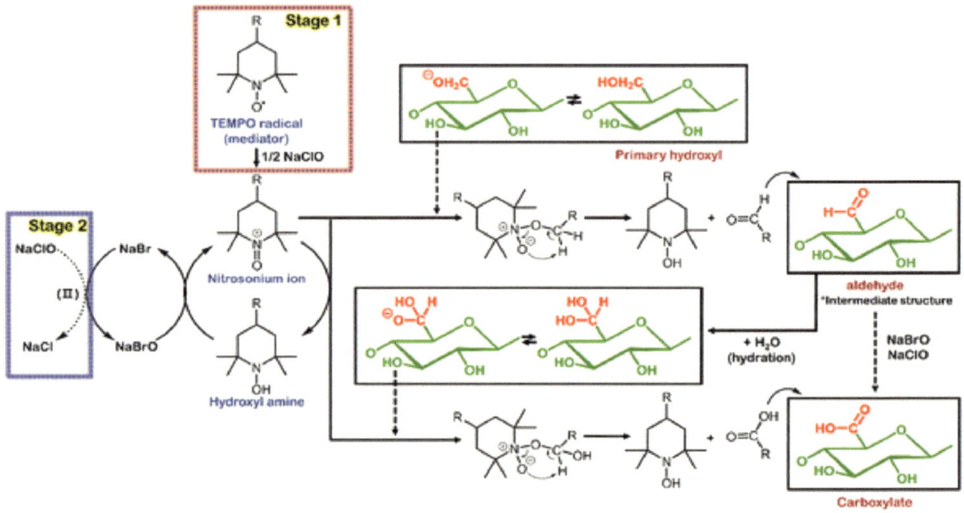

Figure 1.
TEMPO-mediated oxidation scheme. Reproduced from Park et al. 2024 with permission from the Royal Society of Chemistry [20].

subsequent mechanical treatment [26]. On the other hand, parallel side reactions can consume part of the oxidant-producing chlorides, and to a lesser extent chlorates and bromates, that remain attached to the fibers [27], or oxidize hydroxyl groups in C2 and C3 positions forming ketones, although not selectively [28].

Other TMO adaptations have been developed in recent years, including the use of TEMPO derivatives such as 4-amino-TEMPO, which is a more active catalyst than TEMPO [29, 30]; alternative salt combinations that reduce NaBr and, therefore, bromate production [31]; the incorporation of enzymes, like laccases, during the TMO process to promote C6 aldehydes instead of carboxyl groups [29]; or the use of magnetic TEMPO, where a supported TEMPO catalyst prevents its loss in the reaction medium [32]. Nevertheless, these systems remain underdeveloped compared to conventional TMO.

Recently, research has focused on improving the conventional TMO system to enhance the efficiency of the reaction and its scalability. In recent years, particular attention has been given to optimizing the oxidation time, reducing chemical dosages, increasing pulp concentration, and recirculating the reaction medium.

3. Optimization of TEMPO-mediated oxidation time

The first step in evaluating the progress of the TMO reaction is to monitor oxidation over time [33]. **Figure 2** illustrates the TMO reaction of bleached eucalyptus pulp with the TMO conditions established by Saito et al. [21] (referred to as traditional conditions): 10 g of pulp/L reaction medium, 0.1 mmol of TEMPO, and 1 mmol of NaBr per gram of pulp. This reaction is carried out in a stirred reactor using 5 mmol NaClO/g pulp and maintaining the pH at 10 by adding NaOH [21].

The reaction begins with the oxidation of C6 hydroxyl groups within the disordered regions. These groups are, first, transformed into aldehydes and, subsequently, into carboxyl groups, generating electrostatic repulsion between the chains.

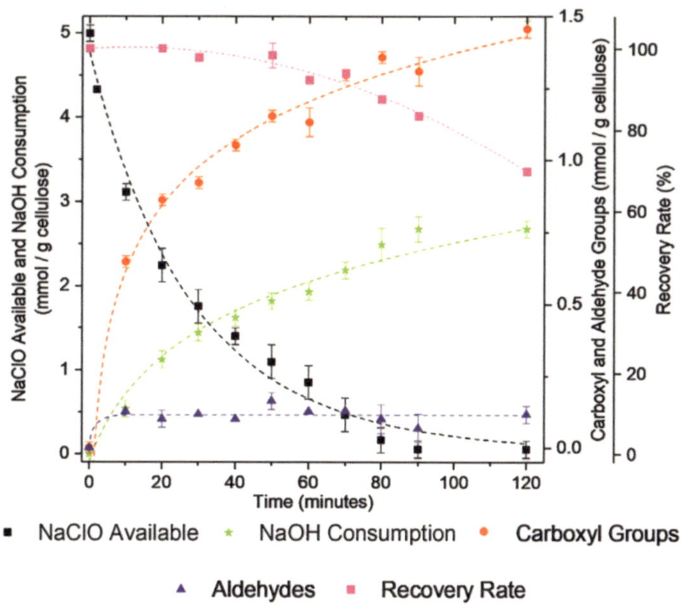

Figure 2.
Monitoring TEMPO-mediated oxidation of bleached eucalyptus pulp using traditional conditions of TMO [33].

The conversion rate from aldehydes to carboxyl is very quick, so the number of aldehydes in the cellulose remains stable over time at less than 0.2 mmol/g of pulp (blue line). Some authors describe this pair of reactions as a consecutive first-order reaction system with respect to the substrate. However, it can be simplified to a single reaction by assuming the aldehyde production rate as the kinetic constant of the overall process [34–36].

At the beginning of the reaction, most of the NaClO available in the medium (black line) is consumed within a few minutes, while carboxyl groups are quickly produced, as shown by the red line. After 10 minutes of reaction, approximately 0.7 mmol/g of carboxyl groups are produced (50%), increasing to around 1.15 mmol/g after 50 minutes. This represents roughly 80% of the carboxyl groups produced after 2 hours (1.45 mmol/g), with the consumption of less than 4 mmol of NaClO per gram of pulp. However, once the pulp reaches a certain oxidation degree, the reaction rate slows down, decreasing the consumption of oxidant and, consequently, the amount of NaOH required to maintain the pH at 10.

Although the OPs did not achieve the same carboxyl content as after 2 hours of oxidation, they contained enough carboxyl groups for the effective production of CNFs after four passes of homogenization at 600 bar with a 1 wt.% consistency. While the cationic demand, transmittance, and nanofibrillation yield of CNFs oxidized for 50 and 120 minutes show similar results, these properties are significantly lower for shorter oxidation times. Additionally, TEM images of CNFs oxidized for 50 or more minutes show a homogeneous diameter range from 25 to 40 nm, without the presence of bigger fibrils [33].

On the other hand, although cellulose oxidation becomes minimal after 50–60 min reaction time, the cellulose recovery rate (pink line) decreases from 90 to 95% at this point to 70% at 120 minutes, when all the NaClO is consumed. This decline is attributed to hydrodynamic conditions and oxidant residues still present in the reaction, which lead to cellulose dissolution as well as to higher losses during the filtration step. Other authors have described bamboo cellulose dissolution as a time-dependent process with two stages: the first one occurs on the surface of the fibers at the beginning of the reaction, causing a minimal cellulose degradation, while the second stage occurs in the inner of the fibers [37]. Among the most important parameters affecting cellulose dissolution, oxidant concentration has the greatest impact. Moreover, by reducing NaClO dose to only 3 mmol/g pulp, the fibers experience only the first stage of the dissolution process, resulting in a very low final pulp dissolution [37]. Therefore, the next step proposed to optimize the traditional TMO process is the evaluation of oxidant doses.

4. Optimization of oxidant consumption

Since the oxidant is necessary for the production of carboxyl groups while it also dissolves cellulose in the medium, a commitment situation is required. Therefore, it is essential to evaluate the oxidant influence on the content of carboxyl groups and aldehydes, as shown in **Figure 3**. The monitoring of NaClO consumption during TMO reaction, using 2.5, 5, and 10 mmol of NaClO under the TMO traditional conditions, is represented in **Figure 3a** [38]. It is observed that the oxidant consumption rate is not affected by the initial NaClO concentration, obtaining a similar consumption rate until all NaClO is depleted. Therefore, the reaction rate can be considered zero-order with respect to NaClO [39]. As observed in **Figure 2**, the consumption rate is elevated during the first minutes of the reaction due to the availability of C6 hydroxyl groups in the disordered regions, whereas progressively, the oxidation of the cellulose surface slows down the reaction rate.

Figure 3b shows the fast stabilization of aldehydes, at around 0.25 mmol/g, regardless of the oxidant dose. In contrast, carboxyl groups keep increasing rapidly, up to 40 minutes, due to the accessibility and feasibility of oxidizing the surface. Then, a second phase of the reaction is observed, in which the amount of carboxyls slightly increases while the oxidant is consumed in other side reactions and chain cleavage. After 30 minutes, more than 1.5 mmol COOH/g pulp is reached using 5 or 10 mmol of oxidant. However, with 2.5 mmol of NaClO/g, this value is not reached,

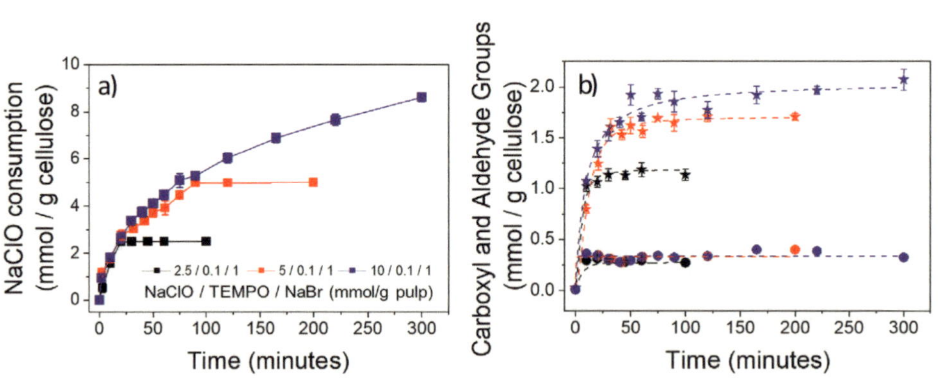

Figure 3.
TMO reaction with different NaClO doses: (a) NaClO consumption and (b) production of carboxyl groups (stars) and aldehydes (circles) [38].

as the NaClO is consumed in less than 25 minutes, obtaining 1.1–1.2 mmol COOH/g without reaching the stabilization phase.

On the other hand, as expected, pulp yield shows a decreasing trend over time, with 2.5 mmol of oxidant providing slightly better yields than the higher doses. Regarding the final properties of CNFs, similar cationic demand, transmittance and appearance in TEM images are observed when using 5 and 10 mmol of NaClO across all reaction times. For 2.5 mmol of oxidant, these properties remain similar at 10 and 20 minutes, while the NaClO is not yet fully consumed.

Therefore, oxidation with 2.5 mmol NaClO for 20–30 minutes may be sufficient to obtain highly fibrillated nanofibers, as 1 mmol COOH/g pulp allows easy nanofiber separation during homogenization. For applications requiring very high fibrillation with no microfibrillated residues, using 5 mmol NaClO/g for 50 minutes would be feasible. However, higher oxidant doses or longer oxidation times are not justified when working with eucalyptus bleached pulp, although these parameters may vary for other types of pulp.

5. Reduction of catalysts dosages in TMO reaction

Another approach to improving the TMO reaction is the adjustment of both catalyst doses, NaBr and TEMPO [38]. The traditional conditions for catalysts in the TMO reaction require 0.1 mmol of TEMPO and 1 mmol of NaBr per gram of pulp, as represented by the blue line in **Figure 4** [21]. In this study, the amount of oxidant was set to 5 mmol/g to ensure that not all of the oxidant would be consumed within at least 1 hour.

Xu et al. [38] proposed optimal catalyst conditions (purple line) of 0.025 mmol/g of TEMPO and 0.5 mmol/g of NaBr. Under these conditions, a minimum of 1 mmol/g of carboxyls was achieved within 50 minutes of reaction, while maintaining at least 80% cellulose recovery. Further reduction of the NaBr co-catalyst was discarded, as

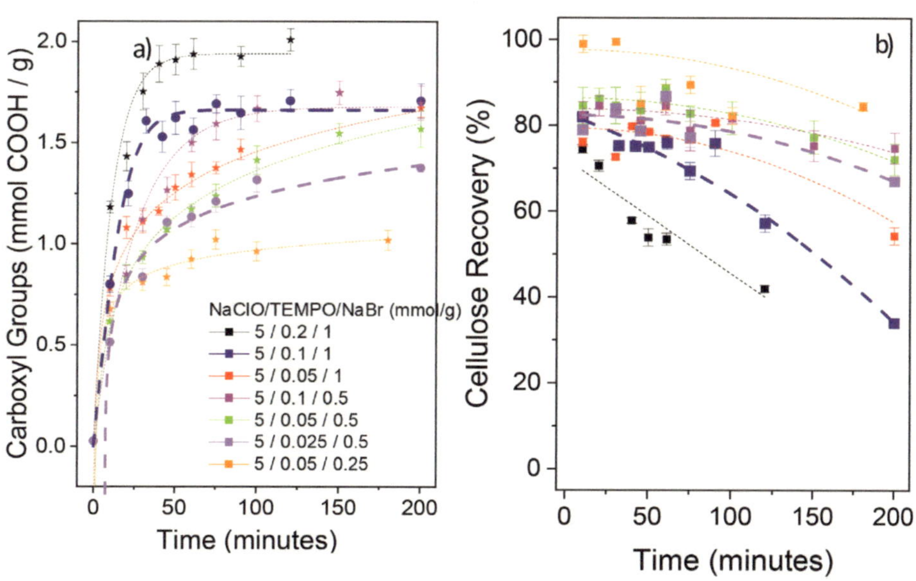

Figure 4.
TMO reaction with different catalyst doses. (a) Carboxyl groups evolution and (b) cellulose recovery [38].

	Optimized conditions of Xu et al. [38]	**Traditional conditions of Saito et al. [21]**
NaClO/TEMPO/NaBr (mmol/g)	5/0.025/0.5	5/0.1/1
Oxidation time (min)	50	120
Carboxyls (mmol/g pulp)	1.1	1.7
Cellulose Recovery (%)	79	57
Diameter (Geometric avg.)	23.1 nm	20.8 nm
Cationic Demand (μeq/g)	1250	2080
Transmittance at 600 nm (%)	60.0	90.6
Polymerization degree (monomers)	256 ± 16	168 ± 5
TEM (x1500)	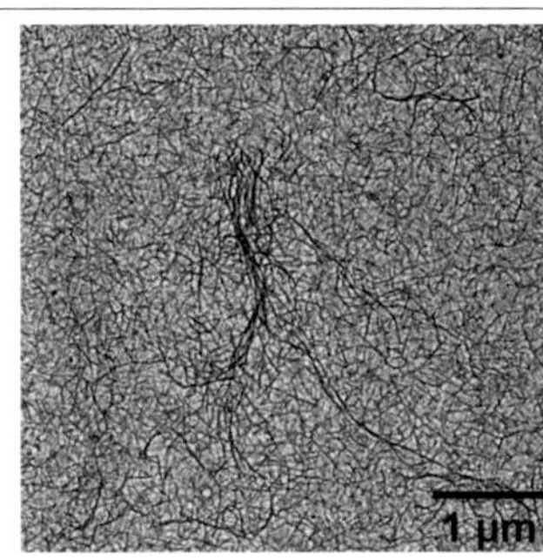	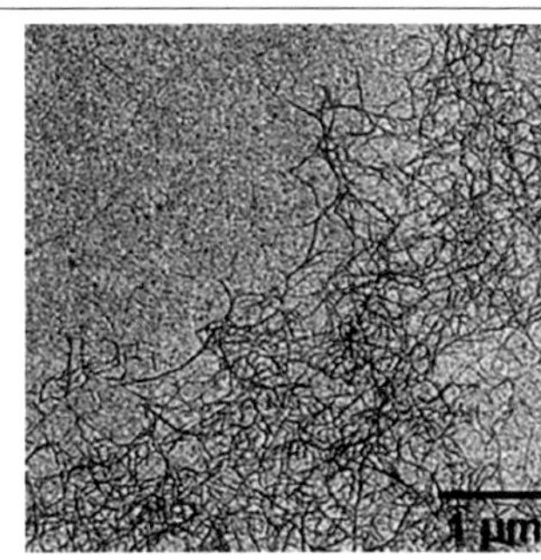

Table 1.
Comparison of CNF properties using the traditional TMO conditions and the optimized conditions proposed by Xu et al. [38].

the number of carboxyl groups was too low (orange line), although cellulose recovery was nearly complete. They suggested reducing NaBr by 50% and TEMPO by 75%. Under these conditions, after 50 minutes of oxidation, the consumption of NaClO was only 2–2.5 mmol/g, indicating that 2.5 mmol/g of NaClO is sufficient. These OPs were used to produce CNFs through 4 passes at 600 bar in the HPH. **Table 1** shows a comparison of the most important properties of CNFs obtained under different TMO conditions. CNFs produced under the optimized conditions exhibit a range of diameters similar to those obtained from more oxidized pulps despite not reaching the same level of transmittance. They also show less breakage of anhydroglucose chains compared to more oxidized pulps, while still achieving sufficient carboxyl groups and higher cellulose recovery during the TMO process.

6. Effect of pulp concentration

Once the catalysts and the oxidant were optimized, another challenge was to increase the pulp concentration in the reactor, set at only 10 g/L. **Figure 5a** shows the evolution of NaOH consumption during the TMO reaction using the optimized catalyst and oxidant doses proposed by Xu et al. [38]. The NaOH consumption follows the same trend as NaClO consumption [40]. It is observed that as pulp concentration increases, the rate of oxidant consumption also accelerates. At a

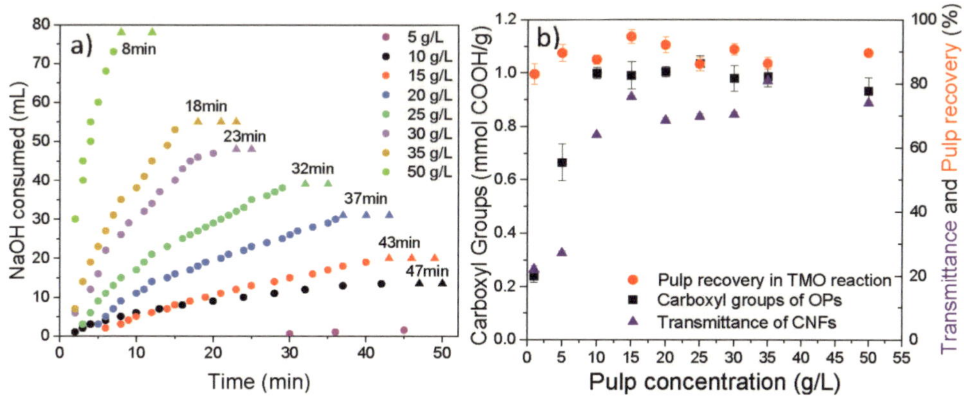

Figure 5.
TMO reaction with different pulp concentrations. (a) TMO reaction evolution with NaOH consumed; (b) carboxyl groups and cellulose recovery in TMO reaction and transmittance of final CNFs [40].

concentration of 10 g/L, the oxidant is fully consumed after 47 minutes, whereas at 50 g/L, it is consumed in only 8 minutes. This observation aligns with findings from other authors, who indicate that the TMO reaction is first-order with respect to pulp concentration [34, 35].

Figure 5b shows the carboxyl group content of OPs and cellulose recovery during the TMO reaction. When the pulp concentration in the reactor exceeds 10 g/L, carboxyl groups reach approximately 1 mmol/g, whereas lower concentrations do not achieve this level, resulting in less efficient oxidation. In all cases, cellulose recovery remains above 80%. Regarding CNF transmittance after homogenization, pulp consistencies above 10 g/L result in transmittance levels of 70–80%. In summary, the reaction can be performed at concentrations higher than 10 g/L, with shorter reaction times, while preserving the CNF properties.

However, the stirring capacity of the reactor must be considered, since at concentrations above 30–35 g/L stirring is difficult, leading to dead zones, with a slight decrease in the carboxyl group content at 50 g/L. Therefore, for concentrations above 30 g/L, alternative stirring systems should be explored, as will be discussed at the end of the chapter.

7. Recycling the reaction medium

According to sustainable principles, the large amount of basic residues generated in the TMO reaction needs to be reduced. Therefore, researchers have focused on the reuse of the reaction medium. **Figure 6** shows the reuse of the reaction medium after oxidation, conducted at 10 g/L pulp consistency under the optimized conditions suggested by Xu et al. [38]. On one hand, since not all the reaction medium is recovered after the pulp oxidation, the amount of cellulose that can be oxidized in subsequent cycles with the recovered medium is reduced unless additional fresh medium is prepared. On the other hand, the incorporation of fresh NaClO into the reaction medium dilutes pulp concentration across cycles (**Figure 6a**). At concentrations below 10 g/L, a gradual reduction in carboxyl groups in the pulp is observed (**Figure 6b**), as shown in the pulp concentration study. Simultaneously, conductivity increases with the number of cycles, and the final properties of CNFs diminish, as

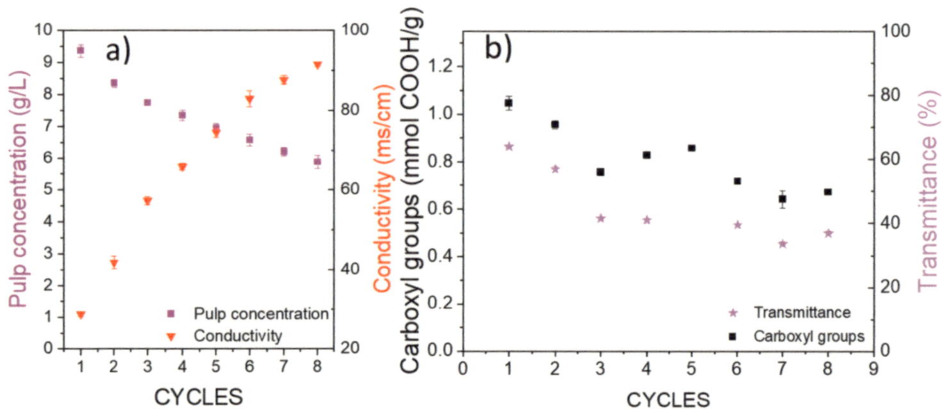

Figure 6.
Reuse of the TEMPO-mediated oxidation medium under optimized conditions: (a) parameters studied for monitoring the reaction and (b) oxidized pulp and CNF characterization (adapted from Xu et al., 2023) [40].

indicated by transmittance levels. These results suggest that salt accumulation or pulp concentration dilution may contribute to the lower CNF quality.

To study the influence of salts, Xu et al. investigated the presence of different salt contents in TMO reaction [40], concluding that the increase of salts in the TMO medium does not alter the amount of carboxyl groups in the OPs. However, the CNF transmittance increases with higher NaCl content. This effect is due to the presence of sodium counterions in the CNFs, which increases fiber swelling and favors nanofibrillation.

Regarding pulp dilution, it was proposed to work with higher pulp concentrations, around 30 g/L, which is the maximum concentration achievable in stirred reactors. To prevent a reduction in the amount of cellulose oxidized in each cycle, the reuse of 75% of the reaction medium from the previous cycle was suggested, with 25% fresh medium added in each cycle [40]. In this 25%, fresh catalysts and the oxidant required for the new cycle are introduced, ensuring no progressive pulp dilution.

Figure 7a shows that incorporating this partial purge tends to reduce and stabilize conductivity over the cycles while maintaining a constant pulp concentration of

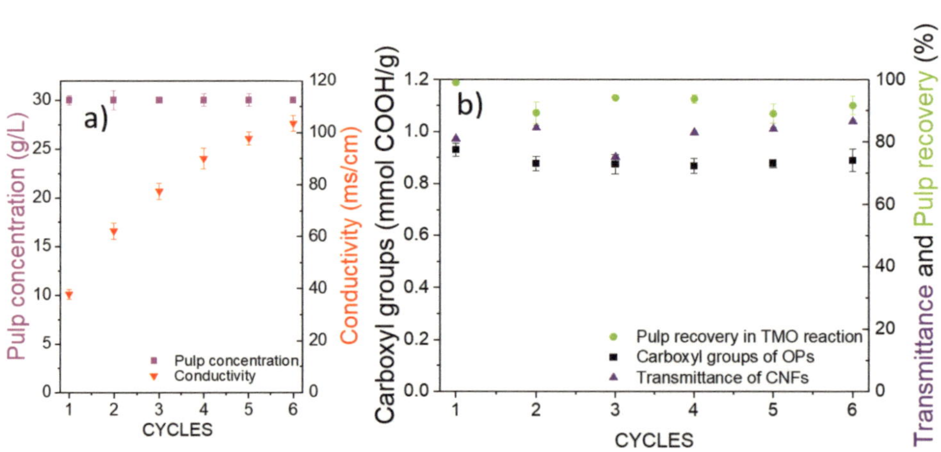

Figure 7.
Reuse of the 75% TEMPO-mediated oxidation medium under optimized conditions: (a) parameters studied for monitoring the reaction and (b) oxidized pulp and CNF characterization (adapted from Xu et al., 2023) [40].

30 g/L in all cycles. Note that the reaction medium in this case contains three times more cellulose than in **Figure 6a**, while the conductivity value remains similar. **Figure 7b** illustrates the preservation of carboxyl groups across oxidation cycles, with pulp recovery of at least 90%. Besides, CNF transmittance is maintained across cycles. These improvements in pulp concentration, pulp recovery, and time (compared to Saito et al. [21]) allow for a more than 1200% increase in pulp production (from approximately 10 to 140 g of OPs/h in a 3 L reactor). Additionally, TEMPO and NaBr consumption are reduced by 94 and 87%, respectively.

8. Steps to move from lab to pilot-plant scale: Kneaders

Despite the remarkable advancements made to render the TMO of cellulose more sustainable and to produce highly fibrillated CNFs, further steps are necessary to scale the process from laboratory scale to pilot-plant trials. Although equipment for stirring, filtration, and homogenization of suspensions is readily available, the main obstacle to scaling up the reaction lies in the low pulp concentrations that can be used in the stirred reactor. To address this limitation, alternative reactor configurations that allow for higher pulp concentrations, such as industrial kneaders or twin-screw extruders, have been explored.

As a proof of concept, Sanchez-Salvador et al. investigated the kneading of small amounts of cellulose in a 500 ml glass beaker at a pulp concentration of 120 g/L [41]. **Figure 8** illustrates the progression of the TMO reaction by monitoring NaOH consumption across different reactor configurations. NaOH consumption provides insight into when the oxidant has been entirely consumed. Similar to the results with higher pulp concentrations in the stirring reactors, kneading pulp at much higher concentrations further reduced reaction times, achieving completion in only 4–5 minutes.

Regarding the carboxyl groups obtained, under the optimized conditions in the stirred reactor, reported by Xu et al. at pulp concentrations of 10 g/L [38] and 30 g/L [40], in both cases approximately 1 mmol/g pulp is reached. In comparison, the kneader used in K-120 conditions produced 0.84 ± 0.04 mmol/g of carboxyl groups. Despite the slightly lower carboxyl group content, it is sufficient to obtain highly

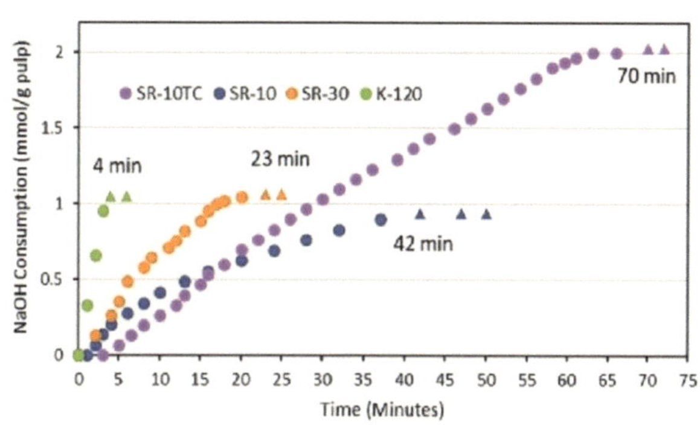

Figure 8.
TMO reaction evolution with NaOH consumed in different reactor configurations [41]. SR-10TC: Stirring Reactor (10 g/L) with Saito et al. conditions [21] and 5 mmol/g of NaClO. SR-10: Stirring Reactor (10 g/L) with Xu et al. optimized conditions [38] and 2.5 mmol/g of NaClO. SR-30: Stirring Reactor with Xu et al. optimized conditions at 30 g/L [40] and 2.5 mmol/g of NaClO. K-120: Kneading at 120 g/L (Proof of concept) with the optimized Xu conditions and 2.5 mmol/g of NaClO.

fibrillated CNFs, with better results in terms of polymerization degree and transmittance compared to stirred reactors at 10 and 30 g/L under the same catalyst and oxidant conditions. Furthermore, cellulose recovery was higher in the kneader process, reaching 94% of the initial pulp [41]. This study also demonstrated the feasibility of using a carbonate-bicarbonate buffer to maintain the pH at 9.5–10, enabling effective control of the pH due to the high reaction rate. As a result, oxidation was achieved within the same reaction time, yielding 0.90 mmol/g of carboxyl groups and 98% cellulose recovery.

Recently, we scaled up this reaction in a 10 L kneader using a carbonate-bicarbonate buffer to control the pH. A preliminary study identified the optimal agitation system as a mixing paddle with a pulp concentration of 100 g/L, which allowed for effective mixing of the pulp without disintegration. In this process, 250 g of eucalyptus bleached pulp was used in the kneading process, although the equipment has the capacity to oxidize over 1 kg of dry pulp. After adding the pulp at 12–13% consistency into the mixing bowl, the process begins by mixing it with the kneading paddle at about 70 rpm for 1 minute to ensure it behaves like a dough. Then, the buffer and catalysts (0.025 mmol of TEMPO/g and 0.5 mmol NaBr/g pulp) are added, stirring for an additional 1–2 minutes. Next, the oxidant (2.5 mmol NaClO/g) is added to achieve the final pulp concentration of 100 g/L. The dough is stirred for 10 minutes to ensure pulp oxidation, although a shorter time could be sufficient. Characterization of the OPs showed carboxyl groups of 1.00 ± 0.05 mmol/g, similar to those obtained in stirred reactors at 10 and 30 g/L, with a pulp recovery of 93%. With the described operation sequence, 250 g of dry pulp per batch can be processed in less than 20 minutes, enabling the production of over 700 g/h of OPs.

9. Steps to move from lab to pilot-plant scale: Twin-screw extruder

Another proposal to improve TMO production is the use of twin-screw extruders (TSE), which allow continuous reactions and enable a soft oxidation of the pulp while simultaneously facilitating fibrillation through the mechanical action of the TSE. Sanchez-Salvador et al. utilized a co-rotating pilot-scale TSE consisting of six modules of 100 mm length (Evolum®25 A110, Clextral, Firminy, France), for the oxidation of eucalyptus bleached pulp [42].

TMO process within the TSE (**Figure 9**) was carried out at room temperature with the TSE speed set at 100 rpm, a residence time of approximately 125 seconds per TSE

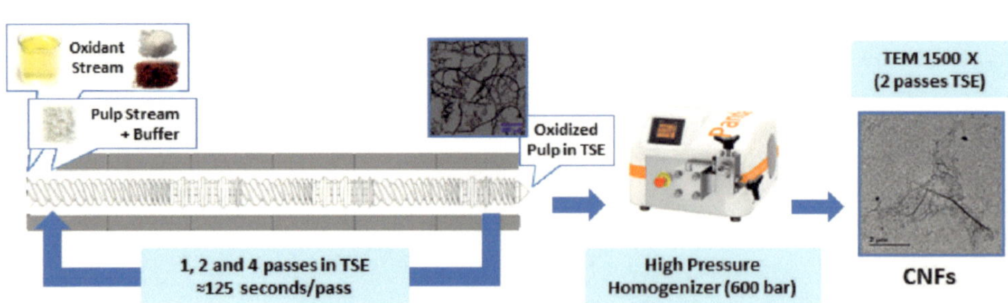

Figure 9.
Scheme of TMO inside TSE and the production of CNFs.

step, and extruding the pulp between 1 and 4 TSE passes. Two streams are introduced into the TSE: one containing disintegrated cellulose pulp and a carbonate-bicarbonate buffer to maintain the pH around 9.5–10, and the other containing the catalysts and the oxidant. Both catalyst and oxidant doses similar to those used by Saito et al. [21] (NaBr/TEMPO/NaClO at approximately 0.8/0.08/2 mmol/g pulp) and Xu et al. [38] (NaBr/TEMPO/NaClO at approximately 0.5/0.025/2 mmol/g pulp) were evaluated. These amounts were added either all at once during a single pass, twice in two passes (half each pass), or in four passes (a quarter per pass). After the TMO reaction, the OPs were washed after resting for 4 hours and subsequently homogenized to obtain CNFs.

Table 2 compares the characteristics of pulps treated with TMO in a stirring reactor, TMO within the TSE, and TSE treatment without TMO. Fibrillation of the pulp is observed after TSE treatment, while fiber rupture and swelling occur with the TMO reaction. When TMO is conducted within the TSE, all three effects – fibrillation, fiber rupture, and swelling – are evident in the OPs. After TSE treatment (without oxidation) followed by HPH, micrographs show a high amount of microfibers, including fiber bundles with branching structures. Nevertheless, after TMO in either the stirred reactor or the TSE, the number of microfibers is significantly reduced, with only a few large microfibers remaining, which are smaller when using the extruder. CNFs are observed in both TMO reactions, as shown in **Figure 9**.

Comparing the different TSE configurations using TMO, carboxyl groups obtained with 0.8/0.08/2 mmol/g decrease from 0.81 mmol COOH/g (one TSE step) to 0.69 mmol/g (four steps), while slightly lower values are observed with 0.5/0.025/2 mmol/g, decreasing from 0.69 mmol COOH/g (one TSE step) to 0.65 mmol COOH/g (four TSE steps). The small differences in carboxyl content between the two catalyst doses suggest that the optimized conditions for the stirred reactor are also effective for TMO in the TSE.

Regarding CNFs, transmittance levels remain between 70 and 80% in all cases, whereas the polymerization degree indicates a reduction in the average monomers of the cellulose chains as the number of passes through the extruder increases. This is corroborated by micrographs, which show a decreasing number of large microfibers

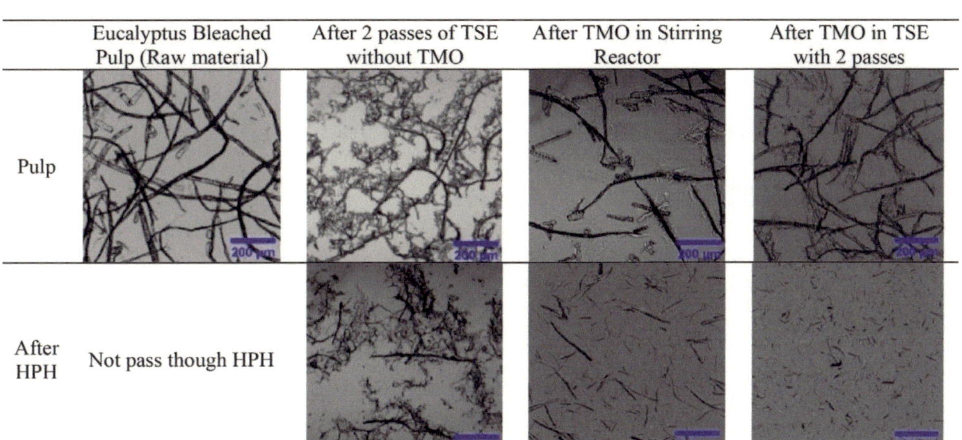

Table 2.
Scheme of TMO inside TSE and the production of CNFs.

with the number of passes in the extruder. Furthermore, the polymerization degree in the TSE samples is lower than that achieved in the stirred reactor, highlighting the fibrillation effect of the extruder.

10. Conclusions and final remarks

The sustainable production of highly fibrillated CNF using the TMO reaction has been demonstrated in recent years with significant advances. Several challenges associated with the reaction have been addressed by monitoring reaction time, reducing oxidant and catalyst doses, increasing pulp concentration, and recycling part of the reaction medium containing the catalysts. Moreover, the exploration of alternative reactor configurations for stirring reactors, such as kneaders (for batch reactions) and twin-screw extruders (for continuous reactions), have been explored. These innovations have facilitated higher oxidized pulp production through the use of higher pulp concentrations, ranging from 100 to 150 g/L, and by accelerating the TMO reaction. Combined with the integration of pilot-plant equipment, these developments support the feasibility of scaling up the TEMPO oxidation process using both batch and continuous reaction systems.

Future efforts to further enhance the scalability and feasibility of CNF production should focus on replacing high-pressure homogenizers, which typically operate with suspensions around 10 g/L, with alternative equipment capable of handling higher pulp concentration. Additionally, exploring novel treatment methods, such as advanced mechanical or chemical techniques, would be a critical next step in improving the efficiency and sustainability of CNF production at an industrial scale.

Acknowledgements

The authors are grateful for the financial support from the Spanish Ministry of Science and Innovation (projects PID2020-113850RB-C21 “CON-FUTURO-ES” and PID2023-147456OB-C22 “NBEA”).

Conflict of interest

The authors declare no conflict of interest.

Author details

Jose Luis Sanchez-Salvador, Hongyu Xu, Ana Balea, Elena Fuente,
M. Concepcion Monte, Angeles Blanco and Carlos Negro*
Department of Chemical Engineering and Materials, Complutense University of Madrid, Madrid, Spain

*Address all correspondence to: cnegro@ucm.es

References

[1] Turbak AF, Snyder FW, Sandberg KR. Microfibrillated cellulose, a new cellulose product: Properties, uses, and commercial potential. Journal of Applied Polymer Science. Applied Polymer Symposium. 1983;**37**:815-827

[2] Wang X, Guo J, Ren H, Jin J, He H, Jin P, et al. Research progress of nanocellulose-based food packaging. Trends in Food Science & Technology. 2023;**143**:104289

[3] Rajinipriya M, Nagalakshmaiah M, Robert M, Elkoun S. Importance of agricultural and industrial waste in the field of nanocellulose and recent industrial developments of wood based nanocellulose: A review. ACS Sustainable Chemistry & Engineering. 2018;**6**(3):2807-2828

[4] Pennells J, Godwin ID, Amiralian N, Martin DJ. Trends in the production of cellulose nanofibers from non-wood sources. Cellulose. 2020;**27**(2):575-593

[5] Bardet R, Bras J. Cellulose nanofibers and their use in paper industry. In: Handbook of Green Materials: Bionanomaterials: Separation Processes, Characterization and Properties. Switzerland: World Scientific; 2014. pp. 207-232

[6] Moohan J, Stewart SA, Espinosa E, Rosal A, Rodríguez A, Larrañeta E, et al. Cellulose nanofibers and other biopolymers for biomedical applications. A review. Applied Sciences. 2020;**10**(1):65

[7] Perumal AB, Nambiar RB, Moses J, Anandharamakrishnan C. Nanocellulose: Recent trends and applications in the food industry. Food Hydrocolloids. 2022;**127**:107484

[8] Chen W, Yu H, Lee S-Y, Wei T, Li J, Fan Z. Nanocellulose: A promising nanomaterial for advanced electrochemical energy storage. Chemical Society Reviews. 2018;**47**(8):2837-2872

[9] Santos RF, Ribeiro JCL, de Carvalho JMF, Magalhães WLE, Pedroti LG, Nalon GH, et al. Nanofibrillated cellulose and its applications in cement-based composites: A review. Construction and Building Materials. 2021;**288**:123122

[10] Peter S, Lyczko N, Gopakumar D, Maria HJ, Nzihou A, Thomas S. Nanocellulose and its derivative materials for energy and environmental applications. Journal of Materials Science. 2022;**57**(13):6835-6880

[11] Lan G-X, Liu Y, Zhou N, Guo D-Q, Ma M-G. Multifunctional nanocellulose-based composites for potential environmental applications. Cellulose. 2023;**30**(1):39-60

[12] D'Acierno F, Michal CA, MacLachlan MJ. Thermal stability of cellulose nanomaterials. Chemical Reviews. 2023;**123**(11):7295-7325

[13] ISO. ISO-TS 20477: 2017 (E): Nanotechnologies: Standard Terms and their Definition for Cellulose Nanomaterial. Switzerland: ISO; 2017

[14] Nechyporchuk O, Belgacem MN, Bras J. Production of cellulose nanofibrils: A review of recent advances. Industrial Crops and Products. 2016;**93**:2-25

[15] Mokhena TC, John MJ. Cellulose nanomaterials: New generation materials for solving global issues. Cellulose. 2020;**27**(3):1149-1194

[16] Qing Y, Sabo R, Zhu J, Agarwal U, Cai Z, Wu Y. A comparative study of cellulose nanofibrils disintegrated via multiple processing approaches. Carbohydrate Polymers. 2013;**97**(1):226-234

[17] Sanchez-Salvador JL, Campano C, Lopez-Exposito P, Tarrés Q, Mutjé P, Delgado-Aguilar M, et al. Enhanced morphological characterization of cellulose nano/microfibers through image skeleton analysis. Nanomaterials. 2021;**11**(8):2077

[18] Isogai A, Hänninen T, Fujisawa S, Saito T. Catalytic oxidation of cellulose with nitroxyl radicals under aqueous conditions. Progress in Polymer Science. 2018;**86**:122-148

[19] Isogai A, Kato Y. Preparation of polyuronic acid from cellulose by TEMPO-mediated oxidation. Cellulose. 1998;**5**(3):153-164

[20] Park J, Lee D, Hwang K, Lee J, Lee T-J, Kim Y, et al. Effect of catalyst and oxidant concentrations in a TEMPO oxidation system on the production of cellulose nanofibers. RSC Advances. 2024;**14**(45):32852-32862

[21] Saito T, Kimura S, Nishiyama Y, Isogai A. Cellulose nanofibers prepared by TEMPO-mediated oxidation of native cellulose. Biomacromolecules. 2007;**8**(8):2485-2491. DOI: 10.1021/bm0703970

[22] Lin C, Zeng T, Wang Q, Huang L, Ni Y, Huang F, et al. Effects of the conditions of the TEMPO/NaBr/NaClO system on carboxyl groups, degree of polymerization, and yield of the oxidized cellulose. BioResources. 2018;**13**(3):5965-5975

[23] Serra A, González I, Oliver-Ortega H, Tarrès Q, Delgado-Aguilar M, Mutjé P. Reducing the amount of catalyst in TEMPO-oxidized cellulose nanofibers: Effect on properties and cost. Polymers. 2017;**9**(11):557

[24] Isogai A. TEMPO-catalyzed oxidation of polysaccharides. Polymer Journal. 2022;**54**(4):387-402

[25] Saito T, Yanagisawa M, Isogai A. TEMPO-mediated oxidation of native cellulose: SEC-MALLS analysis of water-soluble and-insoluble fractions in the oxidized products. Cellulose. 2005;**12**(3):305-315

[26] Saito T, Hirota M, Tamura N, Isogai A. Oxidation of bleached wood pulp by TEMPO/NaClO/NaClO2 system: Effect of the oxidation conditions on carboxylate content and degree of polymerization. Journal of Wood Science. 2010;**56**(3):227-232

[27] Kuutti L, Pajari H, Rovio S, Kokkonen J, Nuopponen M. Chemical recovery in TEMPO oxidation. BioResources. 2016;**11**(3):6050-6061

[28] Hosoya T, Bacher M, Potthast A, Elder T, Rosenau T. Insights into degradation pathways of oxidized anhydroglucose units in cellulose by β-alkoxy-elimination: A combined theoretical and experimental approach. Cellulose. 2018;**25**(7):3797-3814

[29] Jaušovec D, Vogrinčič R, Kokol V. Introduction of aldehyde vs. carboxylic groups to cellulose nanofibers using laccase/TEMPO mediated oxidation. Carbohydrate Polymers. 2015;**116**:74-85

[30] Hirota M, Tamura N, Saito T, Isogai A. Oxidation of regenerated cellulose with NaClO2 catalyzed by TEMPO and NaClO under acid-neutral conditions. Carbohydrate Polymers. 2009;**78**(2):330-335

[31] Inamochi T, Funahashi R, Nakamura Y, Saito T, Isogai A. Effect of coexisting salt on TEMPO-mediated oxidation of wood cellulose for preparation of nanocellulose. Cellulose. 2017;**24**(9):4097-4101

[32] Xu H, Balea A, Merayo N, Martinez A, Negro C. Improving sustainability of cellulose nanofibrils production: FTIR spectroscopy for online control of the synthesis of recyclable magnetic TEMPO catalyst. Carbohydrate Polymer Technologies and Applications. 2024;**7**:100417

[33] Sanchez-Salvador JL, Campano C, Negro C, Monte MC, Blanco A. Increasing the possibilities of TEMPO-mediated oxidation in the production of cellulose nanofibers by reducing the reaction time and reusing the reaction medium. Advanced Sustainable Systems. 2021;**5**(4):2000277

[34] Sun B, Gu CJ, Ma JH, Liang BR. Kinetic study on TEMPO-mediated selective oxidation of regenerated cellulose. Cellulose. 2005;**12**(1):59-66. DOI: DOI 10.1023/B:CELL.0000049409.56806.da

[35] Dai L, Dai H, Yuan Y, Sun X, Zhu Z. Effect of TEMPO oxidation system on kinetic constants of cotton fibers. BioResources. 2011;**6**(3):2619-2631

[36] Mazega A, Santos AF, Aguado R, Tarrés Q, Fiol N, Pèlach MÀ, et al. Kinetic study and real-time monitoring strategy for TEMPO-mediated oxidation of bleached eucalyptus fibers. Cellulose. 2023;**30**(3):1421-1436

[37] He L, Yang S, Guan Q, Li J, Peng L, Zhang J. Kinetic dissolution behavior and mechanism of bamboo cellulose fiber by TEMPO-catalyzed oxidation. Cellulose. 2019;**26**:7089-7097

[38] Xu H, Sanchez-Salvador JL, Balea A, Blanco A, Negro C. Optimization of reagent consumption in TEMPO-mediated oxidation of eucalyptus cellulose to obtain cellulose nanofibers. Cellulose. 2022;**29**(12):6611-6627

[39] Mao L, Ma P, Law K, Daneault C, Fo B. Studies on kinetics and reuse of spent liquor in the TEMPO-mediated selective oxidation of mechanical pulp. Industrial & Engineering Chemistry Research. 2010;**49**(1):113-116

[40] Xu H, Sanchez-Salvador JL, Blanco A, Balea A, Negro C. Recycling of TEMPO-mediated oxidation medium and its effect on nanocellulose properties. Carbohydrate Polymers. 2023;**319**:121168

[41] Sanchez-Salvador JL, Xu H, Balea A, Blanco A, Negro C. Enhancement of the production of TEMPO-mediated oxidation cellulose nanofibrils by kneading. International Journal of Biological Macromolecules. 2024;**261**(2):129612

[42] Sanchez-Salvador JL, Blanco A, Duque A, Negro MJ, Manzanares P, Negro C. Upscaling cellulose oxidation: Integrating TEMPO-mediated oxidation in a pilot-plant twin-screw extruder for cellulose nanofibril production. Carbohydrate Polymer Technologies and Applications. 2024;**7**:100525

Chapter 4

Nanocellulose for Wastewater Treatment

Madhu Sasikala, K. Sivaprakash, M. Lavanya and V. Ramesh Prabu

Abstract

Water pollution poses a serious threat, adversely affecting ecosystems and human health. Addressing this issue is essential to preserve water quality and protect vital resources. Implementing effective solutions is key to safeguarding ecosystems, promoting public health, and ensuring sustainable access to clean water for both present and future generations. The growing global concern over the contamination of natural resources, particularly freshwater, has heightened the demand for effective water treatment methods. There is growing interest in using renewable and eco-friendly reactive materials to support environmental sustainability and combat harmful contaminants. Among the various options, nanocellulose stands out as a promising candidate for environmental remediation due to its remarkable physicochemical properties. This chapter explores the intrinsic characteristics and structural features of different types of nanocellulose, including cellulose nanocrystals (CNCs), cellulose nanofibers (CNFs), and bacterial nanocellulose (BNCs). It has been extensively studied as an excellent biomaterial for wastewater treatment due to its outstanding properties including antifouling behavior, thermal resistance, large specific surface area, superior mechanical strength, high aspect ratio, biodegradability, and biocompatibility. Its extensive surface area is rich in free hydroxyl groups, which are easily modified and functionalized. This chapter also highlights recent advancements in nanocellulose applications, including heavy metal removal, oily water separation, and dye extraction, showcasing its potential as an effective adsorbent for improving filtration performance.

Keywords: adsorption, heavy metal, dye removal, nanocellulose, oil water separation

1. Introduction

Water is one of the most essential resources for sustaining life on Earth. The shortage of clean water to meet human drinking and sanitation needs poses a significant challenge to public health and productivity [1]. This scarcity not only hampers economic development but also undermines efforts to maintain a clean environment and sustain healthy ecosystems. However, the rapid growth of the global population, industrial development, and urbanization have significantly contributed to widespread water pollution [2]. This issue remains a persistent global challenge, making water management a critical component of environmental governance [3, 4].

Wastewater generated from industry and human activities contains a variety of harmful substances, including antibiotics, organic dyes, heavy metal ions, petroleum products, pharmaceuticals, and organic solvents. The long-term accumulation of these pollutants in the environment poses serious threats to human health and other living organisms causing substantial ecological damage.

Hence, it is extremely important to remove the various organic, inorganic, and biological pollutants that enter our water systems [5, 6]. Wastewater treatment processes, such as sorption, membrane filtration, flocculation, catalytic degradation, and disinfection, are extensively used and these processes when performed using well-designed systems are very efficient in remediating the toxic pollutants in water [7]. Various conventional and advanced technologies have been developed to address the issue of wastewater treatment. Traditional methods, including evaporation, oxidation, flotation, electro-precipitation, coagulation, membrane separation, and ion exchange, are widely employed [8, 9] but often fall short in achieving efficient water purification.

Conventional materials used in water treatment processes such as activated carbons and petroleum-based polymers have a large carbon footprint associated with their production and application [10]. The choice of a specific treatment technique largely depends on several factors, such as the type and concentration of waste, the heterogeneity of the effluent, the desired level of purification, and economic considerations [11]. For a sustainable future, it is essential to adopt cost-effective, eco-friendly green alternatives that offer superior performance and a lower carbon footprint [12]. In this regard, the potential of the most abundant biopolymer available on the planet viz., cellulose has been widely explored.

The emerging field of nanotechnology has made remarkable progress in utilizing nanomaterials derived from sustainable natural resources for wastewater contaminant removal [13]. Among these natural materials, cellulose stands out as a highly abundant, sustainable, eco-friendly, and bio-resorbable polymer. Cellulose can be extracted from various lignocellulosic materials such as wood, hemp, bamboo flax, jute, bagasse, or bacterial fermentation [14]. In recent decades, advancements in processing techniques and analytical instrumentation have enabled scientists to explore cellulose at the nanoscale, revealing its remarkable mechanical properties, biodegradability, and versatility properties that are significantly enhanced compared to its macroscopic forms [15]. Nanocellulose materials are categorized into cellulose nanocrystals (CNCs), cellulose nanofiber (CNFs), and bacterial nanocellulose (BNCs), depending on the isolation method. Although they share a similar chemical composition, they differ in particle size, morphology, crystallinity, and certain properties based on their source of extraction [16]. Cellulose nanocrystals, also known as cellulose nanowhiskers, exhibit high strength and are typically produced through acid hydrolysis of cellulose nanofibrils. On the other hand, cellulose nanofibers, also referred to as microfibrillated cellulose, are obtained through mechanical processes rather than acid hydrolysis [17]. These nanofibers are long, flexible, and form interconnected networks. When broken down into nano-sized components, cellulose forms nanocellulose, which measures 100 nanometers or less [18]. The mechanical properties vary depending on whether it originates from the amorphous or crystalline regions of cellulose. Compared to traditional cellulose, nanocellulose offers distinctive chemical and physical properties, including a high surface area, exceptional mechanical strength, and impressive versatility and functionality [19].

This chapter explores recent advancements in the application of nanocellulose-based materials for water and wastewater treatment, highlighting their mechanisms for removing various contaminants. It delves into the removal of metal ions, organic compounds, and other pollutants, as well as their role in desalination processes, providing a comprehensive overview of these innovative systems.

2. Structure and properties of nanocellulose

Cellulose is a natural macromolecule consisting of linear homo polysaccharide chains linked by β-1,4-glycosidic bonds and stands as the most abundant carbohydrate on the planet [20], with an estimated annual production of 10^{10} tons. These chains can aggregate into microfibrils due to extensive intermolecular and intramolecular hydrogen bonding or form crystalline structures when strong hydrogen bonding occurs between hydroxyl functional groups. It consists of nanoscale fibers or crystals with dimensions typically ranging from a few nanometers to several hundred nanometers. Its surface is rich in hydroxyl groups, enabling chemical modifications for enhanced functionality [21]. The material's high crystallinity contributes to its mechanical robustness, while its ability to form transparent, oxygen-resistant films makes it ideal for packaging and electronic applications. The abundance of hydroxyl groups in nanocellulosic materials makes them highly effective for binding various contaminants, including pharmaceuticals, heavy metals, and dyes. Furthermore, its biocompatibility has opened new avenues in the biomedical field for wound dressings, drug delivery systems, and tissue engineering [22, 23]. The combination of these exceptional properties and their renewable origin underscores the growing interest in nanocellulose as a future-oriented material for sustainable technology. The types and properties of nanocellulose are shown in **Table 1**.

The effectiveness of nanocellulose in wastewater treatment is largely attributed to its exceptional physical and chemical properties:

- *High surface area:* Nanoscale dimensions provide a large surface area for adsorption of pollutants.

	Cellulose nanofiber	Cellulose nanocrystals	Bacterial nanocellulose
Treatment	Mechanical	Acid Hydrolysis	Aerobic Bacteria
Morphology	Long chain	rod/needle	Twisted ribbon
Diameter	< 100 nm	1–100 nm	20–100 nm
Tensile Strength	0.8–1 GPa	7–10 GPa	1.5–1.8 GPa
Surface area	52 m^2/g	532 m^2/g	126 m^2/g
Crystallinity	56–65%	72–85%	82–90%
DP	>500	500–15,000	600–10,000
Tensile Modulus	100 GPa	140 GPa	100 GPa

Table 1.
Types and properties of nanocellulose.

- *Abundance of functional groups:* Hydroxyl groups on the nanocellulose surface allow chemical modification to improve selectivity and binding capacity for various contaminants.

- *Mechanical strength:* Its robust mechanical properties enable the formation of stable filtration membranes.

- *Biodegradability:* As a naturally derived material, nanocellulose offers an environmentally friendly alternative to synthetic materials.

2.1 Preparation of nanocellulose

Nanocellulose is prepared through various chemical, mechanical, enzymatic, and biological methods, each influencing its structure and properties [24]. Chemical methods, such as acid hydrolysis, are commonly used to produce cellulose nanocrystals (CNCs) by breaking down the amorphous regions of cellulose fibers, leaving highly crystalline particles [25]. Mechanical techniques, including high-pressure homogenization and ultrasonication, are employed to create cellulose nanofibers (CNFs) through intense shear forces. Enzymatic treatments using cellulase enzymes offer an eco-friendly approach by partially hydrolyzing cellulose for easier fibril extraction. Additionally, bacterial nanocellulose (BNCs) is directly synthesized by cellulose-producing bacteria like *Gluconacetobacter xylinus*, forming highly pure and

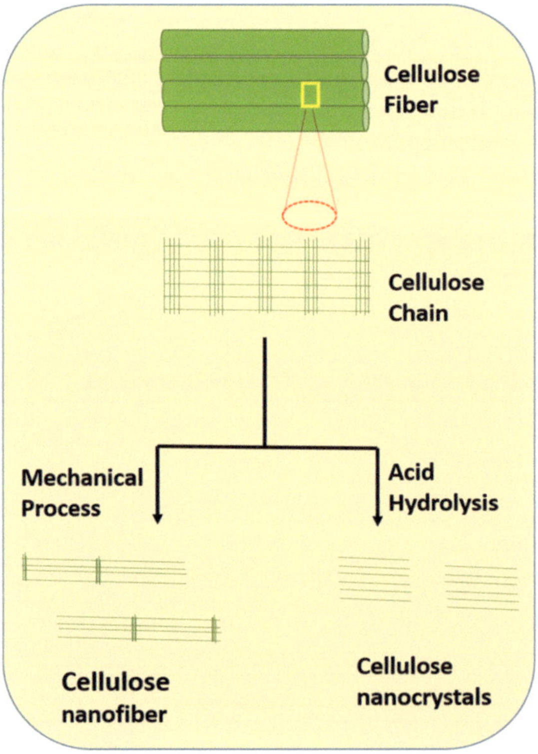

Figure 1.
Structural representation of cellulose.

nanoscale fibril networks [26, 27]. The structural representation of the nanocellulose by acid hydrolysis and mechanical process is depicted in **Figure 1**.

2.1.1 Pretreatment

Pretreatment is necessary to remove lignin from the feedstock, making the carbohydrates accessible for enzymatic hydrolysis and fermentation. Behera et al. [28] state that pretreatment reduces cellulose crystallinity, increases porosity (accessible surface area), and enhances biomass biodegradability by eliminating lignin and hemicellulose. Various pretreatment methods have been proposed to facilitate fiber disintegration [29]. Alkali pretreatment is a widely used chemical process that dissolves lignin in an alkaline solution, making carbohydrates available for downstream processing. Consequently, chemical treatments such as (2,2,6,6-tetramethylpiperidin-1-oxyl) (TEMPO) oxidation, along with enzymatic pretreatment, are commonly applied after alkali treatment to reduce energy consumption in CNF production [30].

2.1.2 Mechanical treatment

A high-intensity mechanical shearing process is employed in a top-down approach to gradually release microfibrils from cellulose pulp fibers, forming cellulose nanofibrils [31]. The process includes grinding, microfluidization, refining, high-shear homogenization, cryo-crushing, and ultrasonication, producing microfibril and nanofibril chains [21]. This treatment creates both crystalline and amorphous domains with a high aspect ratio, exhibiting shear-thinning and leading to gel formation in water [32]. The integration of nanocellulose improved flexural modulus, swelling thickness, and water absorption with increasing nanocellulose content.

2.1.3 Acid hydrolysis

The cellulose molecule consists of both amorphous and crystalline regions, with the amorphous region being removable through acid hydrolysis to yield nanocellulose with high crystallinity. Common hydrolyzing acids include sulfuric, hydrochloric, formic, phosphoric, and hydro-bromic acids. By adjusting factors such as temperature, acid concentration, and hydrolysis duration, nanocellulose with varying sizes and crystallinity can be obtained [33]. A series of processing methods were applied to pineapple peel, producing needle-like nanocellulose with an average width of 14 ± 6 nm and a length of 190 ± 25 nm using ~64% sulfuric acid hydrolysis. This process is adaptable and produces nanocellulose with uniform particle size, with some countries already achieving industrial-scale production.

2.1.4 Oxidative process

The oxidizing agent 2,2,6,6-tetramethylpiperidine-1-oxyl (TEMPO) is relatively weak, but when combined with sodium hypochlorite [34], it forms a nitrogen carbonyl cation—a potent oxidizer that selectively oxidizes C_6 hydroxyl groups on the cellulose surface, converting them into aldehyde and carboxyl groups. TEMPO oxidation offers mild reaction conditions, ease of use, and low energy consumption [35]. However, the process involves hypohalite, which can generate chlorine, posing significant environmental concerns.

2.2 Application of nanocellulose-based wastewater treatment

Nanocellulose has emerged as a highly effective material for wastewater treatment due to its exceptional surface area, mechanical strength, and abundance of hydroxyl groups that can be chemically modified for enhanced functionality [36]. It plays a crucial role in removing various contaminants, including heavy metals, dyes, pharmaceuticals, and organic pollutants. The nanofibrillar structure of CNFs and the high crystallinity of CNCs provide excellent adsorption capabilities, making them ideal for pollutant capture [21]. Moreover, bacterial nanocellulose forms a three-dimensional network that facilitates filtration and adsorption processes. Functionalization of nanocellulose with chemical groups such as carboxyl, amine, or thiol further improves its selectivity and efficiency in contaminant removal [37]. Additionally, nanocellulose-based membranes have shown promise in desalination and advanced filtration applications, offering a sustainable and biodegradable solution for clean water production [38].

3. Methods of contaminant removal

Nanocellulose-based materials remove contaminants from wastewater through several mechanisms:

3.1 Adsorption in dyes

It is a widely used method for water pollutant removal, where pollutants adhere to a solid surface. It is preferred over techniques like polyurethane foams and polymer-based composites due to its simplicity, efficiency, and cost-effectiveness [39]. Hydroxyl groups and chemically modified surfaces bind pollutants such as heavy metals, dyes, and organic compounds.

Activated carbon is a common adsorbent, but its high processing and regeneration costs drive the search for cheaper alternatives. Low-cost adsorbents derived from industrial and agricultural by-products are gaining more interest [40]. Nanocellulose materials, such as CNFs and CNCs, are promising candidates due to their comparable efficiency to high-grade activated carbons, sparking global research interest.

Effluent discharge from industries such as textiles, pulp & paper, dyes, pharmaceuticals, food, and leather tanning contains various dyes that need removal. Nanocellulose-based adsorbents have shown promising dye adsorption capabilities. For example, Sankararamakrishnan et al. synthesized CNF-Fe(0)@FeS from sugarcane bagasse, effectively removing cationic Methylene Blue and anionic Congo Red dyes with adsorption capacities of 200.0 and 111.1 mg/g, respectively [41]. This highlights the potential of agro-waste-derived nanocomposites for industrial dye effluent treatment.

3.2 Filtration

Water filtration membranes eliminate pollutants through size exclusion and charge-mediated adsorption. Nanocellulose materials, with their nanoscale dimensions, high surface area, and mechanical strength, are ideal for filtration applications [42]. It forms membranes with nanoscale pores that physically separate particles and pathogens from water. Depending on pore shape and operation mode, they function as membranes, filters, or adsorbents. Functional groups on its surfaces such as sulfonation, carboxylation, and TEMPO-mediated oxidation enable selective contaminant adsorption. It also reduces bio-fouling due to their hydrophilicity and surface chemistry control [43]. Nanocellulose-based membranes are used in Nano-filtration, ultrafiltration, and membrane distillation, enhancing tensile strength, porosity, and resistance to fouling. Recent innovations, like silver nanoparticle hybrid membranes with antibacterial properties [44], have shown promising applications for bacterial filtration and food packaging.

3.3 Flocculation

Flocculation is a widely used method in municipal and industrial wastewater treatment for removing suspended solids, metals, dyes, and organic matter [45]. Flocculants are introduced to aggregate colloidal particles into larger flocs, which then precipitate. Nanocellulose materials, particularly CNFs, have shown promising results as sustainable flocculants, providing a natural alternative to petroleum-based polymers. Studies demonstrate their effectiveness in reducing wastewater turbidity, though most findings remain at the laboratory level, requiring further research for commercial-scale applications [46].

4. Removal of organic pollutants using catalyst

Water contamination by organic pollutants is a significant and widespread concern. These pollutants encompass a variety of small molecules, including oils, organic dyes, pharmaceuticals, antibiotics, nitro compounds, cosmetic products, and drugs [47]. They can be broadly categorized into water-miscible and immiscible organic pollutants. Immiscible pollutants, such as oils, can be easily removed by separating the organic and aqueous layers. In contrast, water-miscible organic pollutants are more challenging to eliminate [48]. Their treatment typically involves adsorption or advanced degradation techniques, such as UV-visible photocatalysis, ozonation, and oxidation/reduction processes.

Nanocellulose-based materials have been studied for their ability to adsorb various pollutants, including pharmaceuticals, drugs, and antibiotics. One notable approach involved grafting [49] nanocellulose (NC) with a block copolymer, Jeffamine ED 600, denoted as NC–Jeffamine, which effectively adsorbed paracetamol (acetaminophen, ACMP), sulfamethoxazole (SMZ), and N, N-diethyl-meta-toluamide (DEET). The adsorption process was driven by electrostatic interactions. It has also been reported to effectively adsorb the insecticide chlorpyrifos [50]. Both experimental data and computational studies confirmed that nanocellulose efficiently removes chlorpyrifos from aqueous solutions through adsorption.

Dye molecules can be degraded into smaller compounds through various processes, including chemical reactions via reduction, oxidation, and photocatalysis. Chemical degradation relies on redox reactions, which require reducing or oxidizing agents. Cellulose-based materials primarily function as catalysts to enhance these redox reactions. Additionally, they can be combined with photocatalysts such as ZnO, TiO_2, and CuO for degradation under UV and visible light. Cellulose-based membranes have been reported for the adsorption and decolorization of organic dyes, such as methylene blue (MB). They demonstrated adsorption efficiencies of 26, 21, and 35% for CNC, CNC-g-PCysMA, and TOCNF, respectively [51]. However, when used with $NaBH_4$ as a reducing agent, these membranes achieved 100% decolorization of dyes such as RhB. The high decolorization efficiency was attributed to enhanced self-hydrolysis of $NaBH_4$ in the presence of cellulose membranes [52]. Compared to $NaBH_4$ alone, cellulose-based membranes provided 1.5–3 times higher hydrogenation efficiency. Cellulose-MOFs have also been reported as effective catalysts for dye degradation through chemical reduction using $NaBH_4$ [53]. These materials demonstrated high selectivity for anionic dye adsorption, such as MeB, with a 100% removal efficiency. They also acted as catalysts for hydrogenation, enabling the complete degradation of dyes, as confirmed by mass spectrometry. CelloMOF materials can be incorporated into cellulosic filter papers for practical applications.

Metal oxide nanoparticles can spontaneously crystallize on carboxylated cellulose nanofibril-based films, during the adsorption of Cu^{2+} ions from water [54]. This adsorption process facilitates the growth of Cu_2O nanoparticles without requiring additional chemicals or specific experimental conditions like elevated temperature. The simple and efficient formation of Cu_2O nanoparticles on nanocomposite films has enabled their use as photocatalysts for the removal of organic dyes through synergistic adsorption and photocatalytic degradation. The process demonstrated effective dye removal within a short duration. In-situ growth of TiO_2 nanorods (NRs) and gold nanocrystals (NCs) on CNCs has also been demonstrated [55]. The resulting CNC/TiO_2 NRs/Au NCs composite was investigated as a photocatalyst for Rhodium B dye degradation in the presence of H_2O_2, where the plasmonic effect of Au nanoparticles significantly enhanced photocatalytic performance. A CMC/GO/MgO nanoparticle membrane was developed for photocatalytic antifouling applications. This membrane demonstrated the generation of electrons and reactive oxygen species (ROS), making it effective for the oxidation of organic pollutants [56]. Photodegradation of phenol was also achieved using TiO_2 nanoparticles (NPs) embedded in regenerated cellulose (RC). The TiO_2 NPs/RC composite showed superior degradation efficiency compared to individual RC or TiO_2 NPs alone [57].

5. Oil and water separation

Ships transport billions of gallons of petroleum oil annually. In 2020 alone, Organization of the Petroleum Exporting Countries (OPEC) reported a refinery capacity of 12.2 million barrels per day [58]. During transportation, millions of gallons of petroleum oils are spilled into seas and oceans, causing significant harm to aquatic life. Oil floats on water, blocking sunlight and reducing underwater photosynthesis. Various materials, including silica, lignocellulose, zeolites, polypropylene, and natural sorbents, have been used for oil spill clean-up. Cellulose-based materials have been explored as flocculants, adsorbents, and separation membranes for oil/water treatment [59]. Although cellulose is naturally hydrophilic, it can be modified to become a super-hydrophobic material by treating it with hexadecyltrimethoxysilane (HTMS) through chemical vapor deposition (CVD), resulting in aerogels with a water contact angle of 151°. These aerogels demonstrated high adsorption capacities (77–226 g/g) and excellent recyclability over 30 cycles [60]. Notably, they showed an adsorption capacity of 226 g/g for chloroform with high selectivity over various oils and solvents.

6. Conclusion

Nanocellulose-based materials are highly effective in removing contaminants, including heavy metals, dyes, and salts, from wastewater. They are renewable, cost-effective, and sustainable alternatives to traditional remediation materials, thanks to their diverse surface functional groups (carboxyl, hydroxyl, amino, and thiol) that facilitate adsorption and purification processes such as flocculation, membrane filtration, and catalytic degradation. These materials work through mechanisms like hydrogen bonding, electrostatic attraction, and van der Waals forces. However, challenges remain in scaling up for industrial use, including material regeneration and contaminant recovery. Despite these hurdles, nanocellulose holds great potential for future industrial wastewater treatment solutions.

Acknowledgements

The authors express their gratitude to the Department of Research and Development, Chennai Petroleum Corporation Limited, India, for providing the necessary facilities.

Author details

Madhu Sasikala*, K. Sivaprakash, M. Lavanya and V. Ramesh Prabu
Chennai Petroleum Corporation Limited, Chennai, India

*Address all correspondence to: sasi7891@gmail.com

References

[1] Shehu Z, Nyakairu GWA, Tebandeke EO. Overview of African water resources contamination by contaminants of emerging concern. Science of the Total Environment. 2022;**852**:158303. DOI: 10.1016/J.SCITO TENV. 2022. 158303

[2] Aoudi B, Boluk Y, Gamal E-DM. Recent advances and future perspective on nanocellulose-based materials in diverse water treatment applications. Science of the Total Environment. 2022;**843**:156903. DOI: 10.1016/J.SCITO TENV. 2022. 156903

[3] Hendricks DW. Water Treatment Unit Processes: Physical and Chemical. 1st ed. Boca Raton: CRC Press, Taylor and Francis; 2006. 1266 p. DOI: 10.1201/9781315276052

[4] Mohammed N, Grishkewich N, Tam KC. Cellulose nanomaterials: Promising sustainable nanomaterials for application in water/wastewater treatment processes. Environmental Science: Nano. 2018;**5**:623-658. DOI: 10.1039/c7en01029j

[5] Zhang J, Zhong T, Xiang Y, Zhang X, Feng X. Microfibrillated cellulose reinforced poly (vinyl imidazole) cryogels for continuous removal of heavy metals. Journal of Applied Polymer Science. 2021;**138**(48):1-14. DOI: 10.1002/app.51456

[6] Zhang H, Chen Y, Wang S, Ma L, Yu Y, Dai H, et al. Extraction and comparison of cellulose nanocrystals from lemon (Citrus Limon) seeds using sulfuric acid hydrolysis and oxidation methods. Carbohydrate Polymers. 2020;**238**(2):116180. DOI: 10.1016/j.carbpol.2020.116180

[7] Khiari R, Dridi-Dhaouadi S, Aguir C, Mhenni MF. Experimental evaluation of eco-friendly flocculants prepared from date palm rachis. Journal of Environmental Sciences. 2010;**22**:1539-1543. DOI: 10.1016/s1001-0742(09)60286-2

[8] Ali I, Gupta VK. Advances in water treatment by adsorption technology. Nature Protocols. 2007;**1**:2661-2667. DOI: 10.1038/nprot.2006.370

[9] Noor MHM, Azli MFZM, Ngadi N, Inuwa IM, Opotu LA, Mohamed M. Optimization of sonication-assisted synthesis of magnetic Moringa oleifera as an efficient coagulant for palm oil wastewater treatment. Environmental Technology and Innovation. 2022;**25**:102191. DOI: 10.1016/j.eti.2021.102191

[10] Zanaty M, Zaki AH, El-Dek SI, Abdelhamid HN. Zeolitic imidazolate framework@hydrogen titanate nanotubes for efficient adsorption and catalytic oxidation of organic dyes and microplastics. Journal of Environmental Chemical Engineering. 2024;**12**:112547. DOI: 10.1016/j.jece.2024.112547

[11] Tshikovhi A, Mishra SB, Mishra AK. Nanocellulose-based composites for the removal of contaminants from wastewater. International Journal of Biological Macromolecules. 2020;**152**:616-632. DOI: 10.1016/j.ijbiomac.2020.02.221

[12] Peng BL, Dhar N, Liu HL, Tam KC. Canadian Journal of Chemical Engineering. 2011;**89**:1191-1206. DOI: 10.1002/cjce.20554

[13] Herrera-Morales J, Morales K, Ramos D, Ortiz-Quiles EO,

López-Encarnación JM, Nicolau E. Examining the use of nanocellulose composites for the sorption of contaminants of emerging concern: An experimental and computational study. ACS Omega. 2017;**2**:7714-7722. DOI: 10.1021/acsomega.7b01053

[14] Liu Y, Liu H, Shen Z. Nanocellulose based filtration membrane in industrial waste water treatment: A review. Materials. 2021;**14**(18):5398. DOI: 10.3390/ma14185398

[15] Dong Y-D, Zhang H, Zhong G-J, Yao G, Lai B. Cellulose/carbon composites and their applications in water treatment - A review. Chemical Engineering Journal. 2021;**405**:126980. DOI: 10.1016/j.cej.2020.126980

[16] Pandit A, Kumar R. A review on production, characterization and application of bacterial cellulose and its biocomposites. Journal of Polymers and the Environment. 2021;**29**:2738-2755. DOI: 10.1007/s10924-021-02079-5

[17] Kumari P, Raza W, Meena A. Lemongrass derived cellulose nanofibers for controlled release of curcumin and its mechanism of action. Industrial Crops and Products. 2022;**173**(September):114099. DOI: 10.1016/j.indcrop.2021.114099

[18] Moradeeya PG, Kumar MA, Thorat RB, Rathod M, Khambhaty Y, Basha S. Nanocellulose for biosorption of chlorpyrifos from water: Chemometric optimization, kinetics and equilibrium. Cellulose. 2017;**24**:1319-1332. DOI: 10.1007/s10570-017-1197-x

[19] Gopakumar DA, Manna S, Pasquini D, Thomas S, Grohens Y. Nanocellulose: Extraction and application as a sustainable material for wastewater purification. In: New Polymer Nanocomposites for Environmental Remediation. Amsterdam, Netherlands: Elsevier Inc; 2018. 469-486 pp. DOI: 10.1016/B978-0-12-811033-1.00019-6. ISBN: 9780128110331

[20] Tian W, Gao X, Zhang J, Yu J, Zhang J. Cellulose nanosphere: Preparation and applications of the novel nanocellulose. Carbohydrate Polymers. 2022;**277**:118863. DOI: 10.1016/J.CARBPOL.2021.118863

[21] Salama A, Abouzeid R, Leong WS, Jeevanandam J, Samyn P, Dufresne A, et al. Nanocellulose - Based materials for water treatment: Adsorption, photocatalytic degradation, disinfection, antifouling, and nanofiltration. Nanomaterials. 2021;**11**(11):3008. DOI: 10.3390/NANO11113008

[22] Sayyed AJ, Pinjari DV, Sonawane SH, Bhanvase BA, Sheikh J, Sillanpaa M. Cellulose-based nanomaterials for water and wastewater treatments: A review. Journal of Environmental Chemical Engineering. 2021;**9**(6):106626. DOI: 10.1016/J.JECE.2021.106626

[23] Kane AO, Scopel E, Cortez AA, Rossi BR, Pellegrini VOA, Rezende CA, et al. Enzyme-assisted production of cellulose nanofibers from bleached and bleached/sulfonated sugarcane bagasse: Impact of sulfonation on nanocellulose properties and yields. Cellulose. 2023;**30**(18):11507-11520. DOI: 10.1007/S10570-023-05600-2

[24] Kassab Z, Kassem I, Hannache H, Bouhfid R, Qaiss AEK, El Achaby M. Tomato plant residue as new renewable source for cellulose production: Extraction of cellulose nanocrystals with different surface functionalities. Cellulose. 2020;**27**(8):4287-4303. DOI: 10.1007/S10570-020-03097-7

[25] Kaffashsaie E, Yousefi H, Nishino T, Matsumoto T, Mashkour M,

Madhoushi M, et al. Direct conversion of raw wood to TEMPO-oxidized cellulose nanofibers. Carbohydrate Polymers. 2021;**262**:117938. DOI: 10.1016/J.CARBPOL. 2021. 117938

[26] Manian AP, Cordin M, Pham T. Extraction of cellulose fibers from flax and hemp: A review. Cellulose. 2021;**28**(13):8275-8294. DOI: 10.1007/S10570-021-04051-X

[27] Sutrisno E, Tanpichai S, Chuangchote S. Isolation and characterization of cellulose nanofibers (CNFs) from Macaranga hypoleuca. IOP Conference Series: Earth and Environmental Science. 2020;**415**(1):012003. DOI: 10.1088/1755-1315/415/1/012003

[28] Behera S, Arora R, Nandhagopal N, Kumar S. Importance of chemical pretreatment for bioconversion of lignocellulosic biomass. Renewable and Sustainable Energy Reviews. 2014;**36**:91-106. DOI: 10.1016/j.rser.2014.04.047

[29] Abou-Zeid RE, Ali KA, Gawad RMA, Kamal KH, Kamel S, Khiari R. Removal of Cu(II), Pb(II), Mg(II), and Fe(II) by adsorption onto alginate/nanocellulose beads as bio-sorbent. Journal of Renewable Materials. 2021;**9**(4):601-613. DOI: 10.32604/jrm.2021.014005

[30] Reshmy R, Thomas D, Philip E, Paul SA, Madhavan A, Sindhu R, et al. Potential of nanocellulose for wastewater treatment. In: Chemosphere. Elsevier Ltd.; 2021. 130738 p. DOI: 10.1016/j.chemosphere.2021.130738

[31] Pei Y, Wang L, Tang K, Kaplan DL. Biopolymer nanoscale assemblies as building blocks for new materials: A review. Advanced Functional Materials. 2021;**31**(15):1-30. DOI: 10.1002/adfm.202008552

[32] Kumode MMN, Bolzon GIM, Magalh~aes WLE, Kestur SG. Microfibrillated nanocellulose from balsa tree as potential reinforcement in the preparation of 'green' composites with castor seed cake. Journal of Cleaner Production. 2017;**149**:1157-1163. DOI: 10.1016/j.jclepro.2017.02.083

[33] Ibrahim IK, Al-Khateeb Hussin SM, Al-Obaidi YM. Extraction of cellulose nano crystalline from cotton by ultrasonic and its morphological and structural characterization. International Journal of Materials Chemistry and Physics. 2015;**1**(2):99-109. DOI: 10.5772/35482

[34] Chang PS, Robyt JF. Oxidation of primary alcohol groups of naturally occurring polysaccharides with 2, 2, 6, 6-tetramethyl-1- piperidine oxoammonium ion. Journal of Carbohydrate Chemistry. 1996;**15**:819-830

[35] Krause Bierhalz AC. Cellulose nanomaterials in textile applications. Cellulose Chemistry and Technology. 2021;**55**(7-8):725-741. DOI: 10.35812/CELLULOSECHEMTECHNOL. 2021.55.61

[36] Ilyas RA, Sapuan SM, Ishak MR, Zainudin ES. Sugar palm nanofibrillated cellulose (Arenga pinnata (Wurmb.) Merr): Effect of cycles on their yield, physic-chemical, morphological and thermal behavior. International Journal of Biological Macromolecules. 2019;**123**:379-388. DOI: 10.1016/j.ijbiomac.2018.11.124

[37] Arman Alim AA, Othaman R. Epoxidized natural rubber/polyvinyl chloride/microcrystalline cellulose (ENR/PVC/MCC) composite membrane for palm oil mill effluent (POME) treatment. Sains Malaysiana. 2018;**47**(07):1517-1525. DOI: 10.17576/jsm-2018-4707-20

[38] Renu Agarwal M, Singh K. Heavy metal removal from wastewater using various adsorbents: A review. Journal of Water Reuse and Desalination. 2017;**07**(4):387-419. DOI: 10.2166/wrd.2016.104

[39] Sharma V, Shahnaz T, Subbiah S, Narayanasamy S. New insights into the remediation of water pollutants using nanobentonite incorporated nanocellulose chitosan based aerogel. Journal of Polymers and the Environment. 2020;**28**(7):2008-2019. DOI: 10.1007/s10924-020-01740-9

[40] Mo L, Pang H, Lu Y, Li Z, Kang H, Wang M, et al. Woodinspired-nanocellulose-aerogel-adsorbents-with-excellent-selective-pollutants-capture-superfast-adsorption-and-easy-regeneration. Journal-of-Hazardous Materials. 2021;**415**:125612. DOI: 10.1016/j.jhazmat.2021.125612

[41] Alsaiari NS, Katubi KM, Alzahrani FM, Amari A, Osman H, Rebah FB, et al. Synthesis, characterization and application of polypyrrole functionalized nanocellulose for the removal of Cr(vi) from aqueous solution. Polymers. 2021;**13**(21):3691. DOI: 10.3390/polym13213691

[42] Chai F, Wang R, Yan L, Li G, Cai Y, Xi C. Facile fabrication of pH-sensitive nanoparticles based on nanocellulose for fast and efficient As (V) removal. Carbohydrate Polymers. 2019;**245**(December):116511. DOI: 10.1016/j.carbpol.2020.116511

[43] Narkpiban K, Poonsawat T. Optimizing cellulose extraction from kenaf (Hibiscus cannabinus L.) fiber by selective retting and hydrothermal pretreatment. Journal of Natural Fibers. 2020;**00**(00):1-14. DOI: 10.1080/15440478.2020.1758982

[44] Nechyporchuk O, Belgacem MN, Bras J. Production of cellulose nanofibrils: A review of recent advances. Industrial Crops and Products. 2016;**93**:2-25. DOI: 10.1016/j.indcrop.2016.02.016

[45] Tienne LGP, Santos FDPB, de Fatima V, Marques M. Hybrids of cellulose-TiO2 for environmental application. Chemistry and Chemical Technology. 2020;**14**(1):93-101. DOI: 10.23939/chcht14.01.093

[46] Wang K, Liu X, Tan Y, Zhang W, Zhang S, Li J. Two-dimensional membrane and three-dimensional bulk aerogel materials via top-down wood nanotechnology for multi behavioural and reusable oil/water separation. Chemical Engineering Journal. 2019;**371**(April):769-780. DOI: 10.1016/j.cej.2019.04.108

[47] Yuan Z, Wei W, Wen Y. Improving the production of nanofibrillated cellulose from bamboo pulp by the combined cellulase and refining treatment. Journal of Chemical Technology and Biotechnology. 2019;**94**(7):2178-2186. DOI: 10.1002/jctb.5998

[48] Georgouvelas D, Abdelhamid HN, Li J, Edlund U, Mathew AP. All-cellulose functional membranes for water treatment: Adsorption of metal ions and catalytic decolorization of dyes. Carbohydrate Polymers. 2021;**264**:118044. DOI: 10.1016/j.carbpol.2021.118044

[49] Abdelhamid HN. A review on hydrogen generation from the hydrolysis of sodium borohydride. International Journal of Hydrogen Energy. 2021a;**46**:726-765. DOI: 10.1016/j.ijhydene.2020.09.186

[50] Nasser Abdelhamid H, Mathew AP. Cellulose-zeolitic

imidazolate frameworks (CelloZIFs) for multifunctional environmental remediation: Adsorption and catalytic degradation. Chemical Engineering Journal. 2021;**426**:131733. DOI: 10.1016/j.cej.2021.131733

[51] Valencia L, Kumar S, Nomena EM, Salazar-Alvarez G, Mathew AP. In-situ growth of metal oxide nanoparticles on cellulose nanofibrils for dye removal and antimicrobial applications. ACS Applied Nano Materials. 2020;**3**:7172-7181. DOI: 10.1021/acsanm.0c01511

[52] Nair SS, Chen J, Slabon A, Mathew AP. Converting cellulose nanocrystals into photocatalysts by functionalisation with titanium dioxide nanorods and gold nanocrystals. RSC Advances. 2020;**10**:37374-37381. DOI: 10.1039/D0RA05961G

[53] El-Shafai NM, Beltagi AM, Ibrahim MM, Ramadan MS, El-Mehasseb I. Enhancement of the photocurrent and electrochemical properties of the modified nanohybrid composite membrane of cellulose/graphene oxide with magnesium oxide nanoparticle (GO@CMC.MgO) for photocatalytic antifouling and super capacitors applications. Electrochimica Acta. 2021;**392**:138989. DOI: 10.1016/j.electacta.2021.138989

[54] Zeng J, Liu S, Cai J, Zhang L. TiO2 immobilized in cellulose matrix for photocatalytic degradation of phenol under weak UV light irradiation. Journal of Physical Chemistry C. 2021;**114**:7806-7811. DOI: 10.1021/jp1005617

[55] Tan HF, Ooi BS, Leo CP. Future perspectives of nanocellulose-based membrane for water treatment. Journal of Water Process Engineering. 2020;**37**(May):101502. DOI: 10.1016/j.jwpe.2020.101502

[56] Sjahro N, Yunus R, Abdullah LC, Rashid SA, Asis AJ, Akhlisah ZN. Recent advances in the application of cellulose derivatives for removal of contaminants from aquatic environments. Cellulose (Netherlands: Springer). 2021;**28**:7521-7557. DOI: 10.1007/s10570-021-03985-6

[57] Mathivanan Y, Shahir S, Ibrahim Z, Malek NANN. Isolation and characterization of amorphous nanocellulose producing comamonas terrae YSZ sp. from pineapple wastes. Polymer Bulletin. 2024;**81**:15775-15789. DOI: 10.1007/S00289-024-05433-4

[58] Cheng Q, Ye D, Chang C, Zhang L. Facile fabrication of superhydrophilic membranes consisted of fibrous tunicate cellulose nanocrystals for highly efficient oil/water separation. Journal of Membrane Science. 2017;**525**:1-8. DOI: 10.1016/j.memsci.2016.11.084

[59] Fürtauer S, Hassan M, Elsherbiny A, Gabal SA, Mehanny S, Abushammala H. Current status of cellulosic and nanocellulosic materials for oil spill cleanup. Polymers. 2021;**13**(16):2739. DOI: 10.3390/polym13162739

[60] Huang Y, Zhan H, Li D, Tian H, Chang C. Tunicate cellulose nanocrystals modified commercial filter paper for efficient oil/water separation. Journal of Membrane Science. 2019;**591**:117362. DOI: 10.1016/j.memsci.2019.117362

Section 3

Functional Materials Obtained from Agri-Food Residues

Chapter 5

Lignocellulosic Biomass to Produce Functional Materials via DES: Environmental Remediation Applications

Remedios Montenegro, Zoilo González and Alejandro Rodríguez Pascual

Abstract

Due to the abundance, low cost, and inherent properties of the lignocellulosic-based materials, as well as the diverse and scalable synthesis routes and the tunability of the chemical surface, they have been extensively utilized for water and air remediation applications. However, there is a growing trend in research aimed at identifying more sustainable alternatives for the production of these cellulose derivatives. Deep eutectic solvents (DES) have gained significant interest as potential alternatives to traditional solvents due to their versatility, cost-effectiveness, and ability to be designed with multiple components. Their potential as delignification and functionalization agents, as well as their suitability for producing cellulosic micro- and nanofibers, further enhances their appeal in the preparation of cellulose-based materials. This chapter highlights the more recent employment of DES-treated cellulose-based materials used for water and air decontamination treatments through the removal/degradation/reduction of a large range of pollutants. The present study addresses the use of diverse types of DES and their impact on the structural and chemical composition of various (ligno) cellulosic raw materials. Additionally, it explores the potential applications of the cellulose-conformed materials, with an emphasis on nanoscale materials, in remediation processes, highlighting the enhanced outcomes achieved through the utilization of DES.

Keywords: biorefinery, lignocellulose, nanocellulose, deep eutectic solvents, environmental remediation, water remediation, air remediation

1. Introduction

The relationship between human activity and the environment is intimately linked. As the global population continues to expand, there has been a significant rise in energy consumption, the utilization of novel materials, industrialization, the intensity of agricultural activities, and the general movement of the population from rural areas to urban centers [1]. These developments have led to the intensification of environmental pollution, which has emerged as a prominent global concern, significantly impacting both ecosystems and the health of living organisms. The deterioration of the natural environment, which represents one of the most significant challenges currently facing our planet, is a direct consequence of exposure to a multitude of hazardous chemicals, including organic and inorganic substances. This decline in environmental quality has far-reaching implications, not only for human well-being and Earth's ecosystems, but also for the economic security of countries. Among the various affected natural ecosystems, the atmosphere and hydrosphere are of particular concern due to their constant modification by the release and leakage of pollutants. Air pollution can be defined as the introduction of foreign substances in solid, liquid, or gaseous form that results in a reduction in the quality of the air in terms of smell, oxygen content, and health [2]. The primary sources of air pollution are suspended particulate matter ($PM_{2.5}$ and PM_{10}), gases such as NO_2, SO_2, CO_2, CO, and CH_4, and heavy metals [3]. A study conducted jointly by the Health Effects Institute and the Institute for Health Metrics and Evaluation's Global Burden of Disease revealed that in 2021 alone, 8.1 million deaths were attributed to air pollution [4]. In contrast, water pollution occurs when harmful substances enter water bodies without previous treatment [2]. Among the most common pollutants that can be found in water are inorganic anions, heavy metal ions, organic compounds (dyes, detergents, pesticides, and persistent organic pollutants), bacteria, and viruses [5]. As reported by Fuller et al., the premature deaths of a number of people ranging from 1 to 4 million in the year 2019 were attributable to water pollution [6]. Air pollution is considered the leading environmental cause of premature deaths in elderly people, while water pollution has been linked to a considerable incidence of infant mortality [7].

Different physical and chemical treatment methods and technologies have been developed for the abatement of these toxic pollutants. Conventional air remediation technologies comprise thermal/catalytic oxidation [8], plasma and photocatalytic processes [9, 10], condensation [11], and absorption and adsorption processes [12, 13]. However, these processes usually require high energy consumption and high temperatures or pressures [14]. Conventional water treatments, on the other hand, include screening [15], coagulation/flocculation [16], sedimentation [17], filtration [17], oxidation and reduction [18], electrodeposition [19], and adsorption [20]. The majority of these processes are focused on the removal of sediment, sediment-related pollutants, and microorganisms, rather than toxic chemicals [21]. In addition, these technologies present drawbacks such as high energy requirements, capital investment and operating costs, and the production of toxic slurries [22]. The application of biopolymers for water and air treatment has attracted attention due to their convenient availability, biodegradability, and recyclability [23]. Among the various biopolymers, cellulose, the most abundant in the world, is of particular interest due to its distinctive structure and inherent properties [24, 25]. Nanotechnology has recently emerged as a promising alternative to conventional processes, as it allows the tailoring of nanostructured material surfaces through chemical modification or its hybridization with other nanomaterials [26, 27]. Cellulose nanofibers (CNF), obtained through a delamination process of the cellulose fibers, have gained special attention among other nanostructured polymers due to their environmentally friendly nature, high specific surface area, non-toxicity, high mechanical strength, hydrophilicity, and high mechanical strength and tunable surface chemistry [26, 28]. These properties, along with its high adaptability to shape and its abundance of origin, make CNF a perfect candidate to produce materials such as hydrogels, sponges, foams, and aerogels [26],

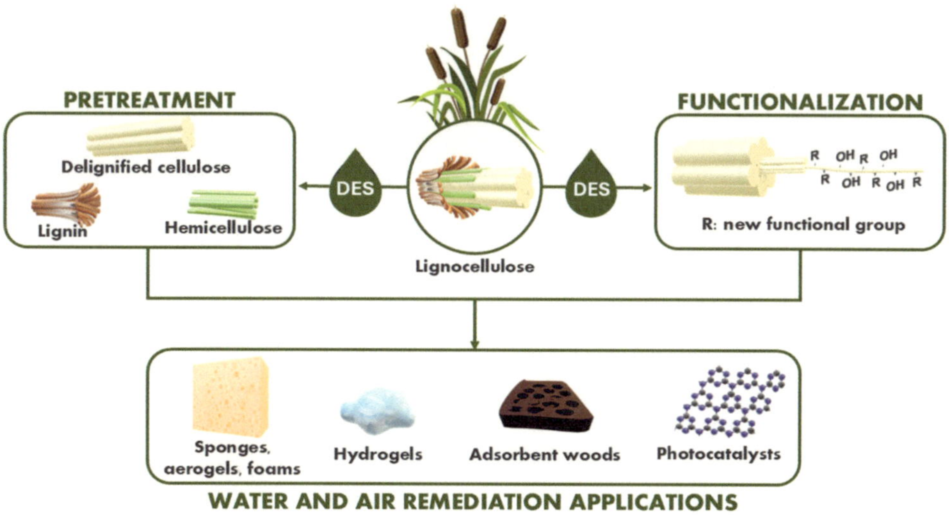

Figure 1.
Application of DES as lignocellulose pretreatment and functionalization strategies for the development of water and air remediation materials.

which can not only adsorb pollutants but also degrade and mitigate them in combination with other photocatalytic nanomaterials.

A substantial number of reviews have already undertaken an exhaustive and satisfactory synthesis of the existing literature on the use of nanocellulose in environmental remediation. The majority of reviews that focus on water purification address the various types of CNFs, their chemical modification, the techniques used for their characterization, and their use as an adsorbent in water treatment [5, 26, 29–35]. The present chapter offers a comprehensive examination of the updated literature about Deep Eutectic Solvent (DES)-pretreated and functionalized lignocellulose materials, with a particular focus on their capacity to remove pollutants from both water and air media, as DES-treated cellulose-based materials' potential in environmental remediation remains largely untapped. This review will initially examine the use of DES for lignocellulose modification and the outcomes of its application. Afterward, it will discuss the use of the aforementioned DES-obtained cellulosic materials in environmental remediation applications, along with the results obtained (**Figure 1**). To the best of our knowledge, this is the first review to evaluate the application of DES-treated cellulose-based materials in air pollution mitigation and purification.

2. Deep eutectic solvents

Deep eutectic solvents have recently emerged as environmentally friendly solvents with a high potential for diverse applications. Despite the various definitions that have been proposed, Martins et al. [36] provided a precise definition of DES as "a mixture of pure compounds for which the eutectic point temperature is below that of an ideal liquid mixture." These mixtures display significant negative deviations from ideality, characterized by a temperature depression that remains above zero [36]. In broad terms, DES are defined as systems that exhibit eutectic properties and are in a liquid state at a specific temperature [37]. These systems often comprise hydrogen bond acceptor interactions (HBA-HBA), hydrogen bond donor interactions (HBD-HBD), and hydrogen bond acceptor and hydrogen bond donor interactions (HBA-HBD) [37]. The classification of DES traditionally comprises five types [38]. Type I is formed by quaternary ammonium salt and a metal chloride, while Type II is constituted by a quaternary ammonium salt and a metal chloride hydrate. The applications of Type I are limited due to the lack of suitable unhydrated metal halides with an adequately low melting point, a challenge that overcomes Type II [39]. Type III is composed of a quaternary ammonium salt acting as an HBA and an HBD. This type is currently the most utilized due to its straightforward synthesis and cost-effectiveness [37, 40]. Type IV is comprised of a metal chloride hydrate with a hydrogen bond donor, while Type V is constituted by nonionic molecular hydrogen bond acceptors and hydrogen bond donors [37, 38]. The use of Type I and II, along with Type IV, is considered toxic because of these metal halides [37].

DES are of particular interest due to their distinctive properties. While the specific characteristics of each DES may vary depending on its source, they are often regarded as environmentally friendly, non-toxic, and biodegradable. Their physicochemical properties, such as a low melting point, low freezing point, high thermal stability, and vapor pressure, make them well-suited to be designed for specific application requirements [40, 41]. Additionally, DES can be recycled, which further enhances their sustainability and cost-effectiveness as solvents [42].

The use of DES has generated considerable interest in the delignification and hemicellulose removal of lignocellulose biomass due to the solvent's ability to effectively fractionate these materials [41, 42]. The use of DES in lignocellulose pretreatment facilitates the penetration of substances from the cell lumen to the cell wall, thereby enabling access to the lignin and hemicellulose present in the secondary wall of the cell and enhancing their solubilization. DES-based pretreatment employs mild conditions, typically not exceeding 175°C, which results in more favorable processes. However, the temperature must be sufficiently high to allow for the accessibility of protons through a reduction in viscosity and surface tension [41]. Additionally, DES have recently emerged as a promising class of functionalization agents for the production of CNF, offering a potential pathway to overcome the limitations of conventional processes, which are often costly and rely on chemicals that are challenging to recycle and reuse [39]. The hydroxyl groups present in the anhydroglucopyranose units of cellulose can be modified through the introduction of functional groups, providing the cellulose with novel functionalities and properties [39, 43].

2.1 DES for lignocellulose pretreatment

A considerable number of DES have been documented for the pretreatment of (ligno)cellulosic samples derived from a large variety of sources [39, 41]. In the context of DES-based pretreatment of materials intended for remediation applications, there is a notable preference for DES composed of choline chloride (ChCl) as an HBA and oxalic acid (OA) or lactic acid (LA) as HBD. This is due to the fact that these compositions exhibit high efficiency in the removal of lignin and hemicellulose while retaining most of the original characteristics of cellulose [44]. The details of the following subsections are presented in **Table 1**, which includes a summary of the chosen raw materials, the used DES, and the selected operational parameters.

Raw material	DES (molar ratio)	Pretreatment conditions	Use of DES	Reference
Balsa wood	ChCl:OA (1:1)	80°C 2 h	Delignification	[44]
Beech wood	ChCl:OA (1:1)	80°C 2, 4, 6, 8 h	Delignification	[45]
Balsa wood	ChCl:OA (1:1)	80°C 2 h	Delignification	[46]
Corncob	ChCl:OA (2:1, 1:1, 1:2, 1:4)	80°C 3 h	Delignification	[47]
Cattail	ChCl:OA (1:1)	50, 70, 90°C 2 h	Delignification, microfibers production	[48]
Commercial cellulose	ChCl:OA:Pr (1:1:2)	80°C 3 h	Cellulose dissolution	[49]
Cypress tree meal	ChCl:OA (2.15:1)	100°C 8 h	Delignification, functionalization	[50]
Almond shell	ChCl:OA (1:1)	200 W 15 min	Delignification, functionalization	[51]
Firmiana simplex bark	ChCl:LA (1:1, 1:2)	100°C, 200°C, 140°C 2, 4, 6 h	Delignification, microfibers production	[52]
Wheat straw	ChCl:LA (1:15)	90°C 12 h	Delignification	[53]
Wood sawdust	ChCl:LA (1:5)	70, 90, 100, 110, 121°C 0.5, 1, 2, 4, 6 h	Delignification	[54]
Beech wood	ChCl:LA (1:2)	120°C 3 h	Delignification	[55]
Palm empty fruit bunches	ChCl:LA ChCl:Bu ChCl:OA (1:10, 1:2, 1:1)	60, 80, 100°C 8 h	Delignification	[56]
Hardwood pulp	ChCl:Glu (2:1)	90°C 5, 10, 20, 30 min	Cellulose dissolution, microfibers production	[57]
Recycled boxboard and recycled milk-container board	ChCl:Ur (1:1.6)	100°C 2 h	Nanocellulose production	[58]
Corn stalk pith	ChCl:pTA (1:1)	100°C 2 h	Delignification	[59]
Residue of hydrolyzation of corncob	MTAC:AA:AlCh (1:1:0.05)	70, 90, 110, 230°C 3 h	Delignification, functionalization	[60]

Table 1.
Different DES-based pretreatments with their operational conditions.

2.1.1 ChCl:OA-based DES for lignocellulose pretreatment

Yang et al. [44] conducted a study to investigate the effect of delignification on balsa wood through three different strategies: the use of $NaOH\text{-}Na_2SO_3$, $NaClO_2$, and a DES composed of ChCl and OA in a 1:1 molar ratio. The findings of the study indicated that, although the DES-pretreatment exhibited a lower efficiency

(60% removal of lignin compared to 72% and 77% for NaOH-Na_2SO_3 and $NaClO_2$, respectively), it demonstrated superior preservation of the wood cell structure and resulted in a cellulose-based wood with a highly porous structure, which was advantageous for the subsequent adsorption application [44]. Zheng et al. [45] similarly observed a similar behavior when they treated beech wood with this DES. The partial removal of lignin and hemicellulose resulted in an organized structure with an improvement in the porosity of the wood, which in turn enhanced the specific surface area and water permeability. Additionally, the incorporation of silver nanoparticles into the porous structure of the resulting wood facilitated the exposure of the nanoparticles to the organic pollutants to be removed catalytically. However, it was observed that an excessive removal of lignin and hemicellulose could result in a reduction of *in situ* growth of the nanoparticles in the channels, thereby inhibiting the catalytic performance of water pollutants [45]. Following the same DES-based pretreatment, Wu et al. [46] delignified balsa wood slices, resulting in the preservation of a lignocellulosic framework with high porosity. This framework was then functionalized through the impregnation of sodium carboxymethylcellulose and chitosan quaternary ammonium salt, ultimately achieving a filter membrane for oil separation [46]. In order to produce cellulose nanofibers, Ren et al. [47] described the use of the ChCl:OA DES with different molar ratios for the extraction of them through a subsequent ultrasonic-assisted treatment of corncobs. In addition to the effective removal of lignin, the treatment was observed to prepare nanofibers with a high crystallinity index, due to the hydrolysis of the amorphous region promoted by oxalic acid, and to enhance the dispersion stability of the suspension. The combined process, followed by a combination of the CNF with alkali lignin, resulted in the preparation of robust aerogels. Subsequent silanization treatment with trimethoxymethylsilane for hydrophobization led to the production of absorbents capable of absorbing various oils and organic solvents [47]. In a related study, cattail pulp, previously obtained through a hydrothermal treatment with NaOH, was further processed using a ChCl:OA treatment to produce microfibers. These were then subjected to the same silanization treatment, after which the resulting solid material was tested as an oil adsorbent [48]. In their work, Wen et al. [49] described the preparation of a ternary DES comprising choline chloride, oxalic acid, and 1,2-propanediol (Pr) at different molar ratios to completely dissolve commercial cellulose. Once the cellulose had been dissolved, it was combined with two additional biopolymers, chitosan and alginate, using polyvinyl alcohol and glutaraldehyde as porogens and cross-linking agents to generate highly effective chitosan-alginate-cellulose aerogels that were capable of adsorbing several anionic dyes [49]. In addition to its use as a pretreatment, Wang et al. [50] and Omer and Fakhre [51] also employed this DES as a functionalization treatment. In their study, Wang et al. [50] subjected cypress tree meal to a process extraction to obtain lignin. The residue resulting from this process, which was composed mainly of cellulose fibers, was reported to undergo carboxylation or esterification reactions as a result of the hydroxyl groups present being subjected to carboxylation and esterification due to the presence of a carboxylic acid. As a consequence of these new functionalities, the cellulose residue was found to become a highly effective adsorbent for neutral red dye [50]. Similarly, Omer and Fakhre observed a comparable effect when they delignified almond shells to obtain cellulose microfibers using a microwave-assisted pretreatment for their use as Pb^{2+} adsorbent [51].

2.1.2 ChCl:LA-based DES for lignocellulose pretreatment

As previously stated, lactic acid is one of the most used hydrogen bond donors for lignocellulose pretreatment. Farooq et al. [52] prepared a DES composed of ChCl:LA, varying the molar ratio to utilize it for the delignification and microfibers' extraction of *Firmiana simplex*. The extracted microfibers were then combined with polyamideamine-epichlorohydrin resin, which acted as a crosslinker, and freeze-dried so as to obtain aerogels that were later employed in air purification applications [52]. Guo et al. [53] applied this same DES in order to utilize both the lignin and the cellulose derived from wheat straw. The ChCl:LA DES enabled the extraction of lignin with high structural integrity, while the resulting cellulose was then loaded with precursors of a bimetallic oxide ($MgFe_2O_4$) and subsequently pyrolyzed. The resultant cellulose-based catalytic biochar was employed for the activation of persulfate for the oxidative degradation of a pharmaceutical pollutant [53]. The ChCl:LA DES was also used by Meas et al. [54] in the delignification of wood sawdust. The resulting pulp was successfully bleached and carboxymethylated. This carboxymethylated cellulose was then combined with sodium alginate to generate spheres capable of adsorbing dyes [54]. Similar to the previously reported works by Yang et al. [44], Zheng et al. [45], and Wu et al. [46], He et al. [55] also demonstrated the efficacy of removing lignin and preserving the original cellulose structure when utilizing a DES system with beech wood, despite the DES being formed by ChCl:LA instead of ChCl:OA. The delignification process enabled the generation of a greater number of binding sites. These were subsequently modified by grafting carboxyl and sulfhydryl groups onto the cellulose surface, thus facilitating the adsorption of Cu^{2+} and Cd^{2+} ions [55]. In a recent study, Khan et al. [56] tested three distinct DES at varying ratios on empty fruit bunches with the objective of removing lignin and producing cellulose nanofibers. This was achieved through a combined ultrasonication and high-pressure homogenization process. The DES were composed of choline chloride as a hydrogen bond acceptor and lactic acid, 1,3-butanediol (Bu), and oxalic acid as hydrogen bond donors. The authors indicated that the highest cellulose yield was 38.78% when ChCl:Bu was utilized. Nevertheless, the highest yield of nanocellulose was observed in the ChCl:LA system (68.49%). The enhanced capacity to interact with cellulose can be attributed to the presence of both carboxyl and hydroxyl groups in lactic acid, which facilitates the disruption of the hydrogen bonding network in cellulose, thereby enhancing its conversion to nanocellulose. Following the production of nanocellulose, tert-butyl alcohol was added prior to freeze-drying to induce pore generation. The resulting filters were employed in the filtration of airborne particulate matter [56].

2.1.3 Other DES for lignocellulose pretreatment

Beyond the frequently studied ChCl:LA and ChCl:OA DES, various other compounds have also been explored as HBAs and HBDs for the pretreatment of lignocellulosic biomass. To modify the micromorphology of cellulose-based carbon aerogels through their surface roughness and hydrophobicity, Ma et al. [57] subjected hardwood pulp to a dissolution treatment using a DES composed of choline chloride and glucose (Glu) in a 2:1 ratio, with the additional inclusion of N-methyl morpholine-N-oxide as a co-solvent. The combination of these three components is capable of forming hydrogen bonds with the hydroxyl groups of cellulose, which in turn disrupts the aggregation observed in microfibers. Following the addition of a cross-linking agent, aerogels were prepared by freeze-drying the suspensions. Some of these were subsequently subjected to a carbonization process. The resulting materials were employed in oil/water separation and organic solvent separation applications [57]. A choline chloride-urea (Ur) DES in a 1:1.6 molar ratio was prepared by Laitinen et al. [58] for an effective DES-based nanofibrillation of cellulose derived from recycled boxboard and recycled mill-container board. For hydrophobization of the final aerogels, CNF suspensions were silylated and freeze-dried. The resulting sponges were then applied in oil/water and organic solvent separations [58]. Peng et al. [59] also prepared hydrophobic aerogels for oil/water separation using silylated cellulose aerogels. However, they employed a DES composed of choline chloride and p-toluenesulfonic acid monohydrate (pTA) for delignification of corn stalk pith, observing the aforementioned well-aligned cellulose skeleton and capillary structure. This ChCl:pTA pretreatment was essential for the effective subsequent oxidation using TEMPO/NaClO/$NaClO_2$. Once the aerogels were obtained, they were subjected to the silanization process referenced previously for the purpose of enhancing their capacity for oil adsorption from water [59]. Recently, Wang et al. [60] presented a novel and circular process for the valorization of corncob residues resulting from the hydrolysis of corncobs. A novel functional ternary DES was prepared, comprising methyl acryloyloxyethyl trimethylammonium chloride (MTAC), acrylic acid (AA), and aluminum chloride (AlCh) in a 1:1:0.05 molar ratio. This was used to pretreat and functionalize the cellulose-rich fraction in the residue with carboxyl groups. The DES was regenerated and reused until its delignification rate was less than 80% when it was converted to a flocculant through self-catalytic *in situ* polymerization. The modified lignin was employed to generate a biochar-based aerogel, and the carboxylated cellulose was then subjected to high-pressure homogenization to produce lignocellulose nanofibers. These fibers were combined with urea and underwent thermal treatment to create a carbon-doped g-C_3N_4, which was subsequently utilized in the photocatalytic degradation of a dye [60].

2.2 DES for lignocellulose functionalization

This section presents an updated and unique review of lignocellulose chemical surface modification with DES for its potential application in water and air remediation. A summary of the following DES-based functionalization studies, along with the selected reaction conditions, can be found in **Table 2**. As previously stated in the earlier section, some authors have employed DES as agents for both the delignification and functionalization of their cellulosic raw materials [50, 51, 60]. However, most authors have focused their attention on applying DES for functionalization exclusively to samples that have already undergone pretreatment or delignification processes, thereby focusing only on the functionalization process. These functionalization or derivatizing treatments of cellulose, particularly nanocellulose, with DES have been recently discussed in depth by Almeida et al., pioneers in highlighting and summarizing the potential of this approach [39].

Selkälä et al. [61] obtained anionic nanofibers from a commercial softwood pulp by subjecting the cellulose to a succinylation reaction through a DES composed of urea and lithium chloride (LiCl) in a molar ratio of 5:1 and using succinic anhydride as the reactive agent. The employment of this DES facilitated the appropriate incorporation of carboxylic acid functionalities into the cellulose backbone, thereby enhancing the production of nanofibers. Subsequently, the hydrogels assembled from these nanofibers were applied as adsorbents for pharmaceutical pollutants [61].

Raw material	DES (molar ratio)	Reactive agent	Reaction conditions	Functionalization	Reference
Cypress tree meal	ChCl:OA (2.15:1)	DES	100°C 8 h	Carboxylation, esterification	[50]
Almond shell	ChCl:OA (1:1)	DES	200 W 15 min	Carboxylation, esterification	[51]
Corncob hydrolyzate residue	MTAC:AA:AlCh (1:1:0.05)	DES	70, 90, 110, 230°C 3 h	Carboxylation	[60]
Commercial softwood pulp	LiCl:Ur (1:5)	Succinic anhydride	70–90°C 2–6 h	Succinylation	[61]
Sugarcane bagasse	TEAOH: 1,3-DMU (2:1)	Glycidyltrimethylammonium chloride	25°C 8 h	Quaternization	[62]
Commercial cellulose acetate monolith	CChCl:Ur (1:2)	DES	95°C 8 h	Quaternization	[63]
Commercial cellulose	AH:Gl (1:2)	DES	80°C 10 min	Aminoguanidine functionalization	[64]
Commercial cellulose	AH:Gl (1:2)	DES	80°C 10 min	Aminoguanidine functionalization	[65]
Spruce wood pulp, sawdust	SA:Ur (1:2)	DES	150°C 30 min	Sulfation, carbamation	[66]
Wheat straw	SA:Ur (1:2)	DES	150°C 30 min	Sulfation, carbamation	[67]
Poplar pulp	SA:Ur (1:2)	DES	150°C 2 h	Sulfation, carbamation	[68]
Poplar pulp	SA:Ur (1:2)	DES	150°C 2 h	Sulfation, carbamation	[69]
Low-value wood	SA:Ur (1:2)	DES	130°C 3 h	Sulfation, carbamation	[70]
Low-value wood	SA:Ur (1:2)	DES	130°C 3 h	Sulfation, carbamation	[70]

Table 2.
Different DES-based functionalization treatments with their operational conditions.

Similarly, Yao et al. [62] employed a DES combined with a reactive agent to enhance the chemical modification of cellulose and promote its nanofibrillation. Specifically, they employed a DES comprising aqueous tetraalkylammonium hydroxide (TEAOH) and 1,3-dimethylurea (1,3-DMU) in a 2:1 molar ratio to facilitate the deprotonation of the hydroxyl groups in sugarcane bagasse, thereby enabling the reaction with glycidyltrimethylammonium chloride as a cationization agent. The resulting nanofibers were employed as hydrogels for the adsorption of dissolved and colloidal substances as a consequence of the presence of quaternary ammonium groups in the nanofibers [62].

Yang et al. [63] subjected a previously treated cellulose monolith directly to a cationization process. The solvent and reactive agent for this process was a DES formed with chlorocholine chloride (CChCl) and urea at a 1:2 molar ratio. Trimethylammonium groups were introduced to the cellulose monoliths, which were then employed as innovative adsorbents for dyes [63].

In the study conducted by Selkälä et al. [64], prior-prepared dialdehyde cellulose, obtained from commercial cellulose, was subjected to a cationization process through the use of an aminoguanidine hydrochloride (AH): glycerol (Gl) DES. Following a mechanical disintegration process, cationic nanofibers were obtained, which were then applied in a sequential removal process of anionic dye chrome azurol S from water. Initially, tubular halloysite nanotubes interacted with the aqueous solution, and subsequently, the CNFs were introduced to promote the aggregation of the dye-loaded nanotubes, with the additional role of contributing to the dye adsorption [64]. Following the same DES-based functionalization process, Selkälä et al. [65] used the DES with the objective of chemically modifying the prior-functionalized cellulose and disintegrating it into nanocrystals. The resulting nanocrystals were combined with a geopolymer to create CNC-geopolymer foam hybrids, which were subsequently employed for the removal of anionic dyes [65].

The employment of the sulfamic acid:urea (SA:Ur) DES has been increasingly prevalent in recent years, with a particular focus on the functionalization of cellulose. This process primarily introduces sulfate groups into the cellulose while simultaneously promoting effective nanofibrillation yields. In a study conducted by Sirviö and Visanko [66], spruce wood pulp and sawdust were effectively functionalized using SA:Ur in a 1:2 molar ratio. The resulting sulfated lignocellulose nanofibers were then subjected to a high-pressure homogenization process, after which they were applied in the simultaneous removal of Pb^{2+} and Cu^{2+} ions as hydrogels. It was observed that the introduction of sulfate groups improved the adsorption of ions and that carbamate groups, which were a consequence of a secondary reaction between cellulose and urea, also contributed to this process [66]. In a similar approach, Montenegro et al. [67] utilized wheat straw as a source of agricultural residue and employed a novel strategy for sulfated lignocellulose and cellulose nanofiber production. These were then utilized in the adsorption of Cu^{2+} ions, both in the form of hydrogel and aerogel, with the aim of studying the difference between the formats and the adsorption kinetics [67]. Zhou et al. [68] introduced sulfate groups into poplar alkaline peroxide mechanical pulp using this DES-based functionalization process for 2 hours with the objective of obtaining sulfated lignocellulose nanofibers. Subsequently, the nanofibers were dispersed with $Bi(NO_3)_3{\cdot}5H_2O$ and KBr to produce BiOBr/sulfated lignocellulose nanofiber composites via *in situ* synthesis. The incorporation of the nanofibers resulted in an increase in the number of oxygen vacancies in BiOBr, thereby enhancing visible light absorption and reducing the recombination of photogenerated carriers. These composites exhibited excellent performance in the adsorption-photocatalytic synergistic removal of organic pollutants in water [68]. The application of the SA:Ur DES for the sulfation and nanofibrillation of bleached low-value wood was studied by Sun et al. [69]. The resulting lignocellulose fibers were subsequently mixed with tourmaline particles with the objective of obtaining hybrid tourmaline-lignocellulose hydrogels. Following a freeze-drying process, the hybrid aerogels were produced and found to have the capacity to produce negative oxygen ions (NOIs), thereby demonstrating the potential for promoting air purification [69]. As a means to improve their own strategy, Sun et al. [70] proposed the use of these DES-functionalized sulfated nanofibers for the production of aerogels containing both tourmaline particles with metal-organic frameworks. These hybrid composites, capable of producing NOIs, were applied in CO_2 adsorption and smoke suppression experiments [70].

3. Environmental remediation applications

3.1 Water remediation

The use of cellulose and nanocellulose fibers as pollutant sorbents in water has been the subject of extensive research [24]. However, the application of DES-pretreated and DES-modified cellulose nanofibers in this field is still in its early stages of development. In addition to the direct use of these fibers for adsorption and absorption, a promising strategy is the combination of these capacities with the synergistic photocatalytic activity of a photocatalyst.

3.1.1 Oil/water separation

Despite cellulose's inherent hydrophilic nature, its tailoring properties have enabled researchers to confer it with hydrophobic characteristics, thereby facilitating its utilization as an oil adsorbent and absorbent.

Laitinen et al. [58] combined the nanocellulose obtained through a DES-pretreatment with two silane solutions in order to achieve silylated CNF sponges with high hydrophobicity and porosity after a freeze-frying process. The sponges were subsequently evaluated for their ability to absorb various oils (marine diesel oil, kerosene, gasoline, motor oil, castor oil, and linseed oil) and organic solvents (dimethyl sulfoxide, chloroform, n-hexane, toluene, acetone, and ethanol). The results demonstrated ultrahigh absorption capacities for all the oils and solvents, with chloroform uptake values reaching 205 g/g and marine diesel oil absorption capacities ranging from 88 to 143 g/g. These values were significantly higher than those of the commercial materials used as a reference. Additionally, the sponges demonstrated the ability to be reused for up to 30 cycles without a notable decline in their absorption capacities [58].

A different strategy was followed by Yang et al. [44], who utilized hexadecyltrimethoxysilane to endow a superhydrophobic delignified surface. The absorptive capacity of the absorbent wood was evaluated with silicone oil, engine oil, dichloromethane, and n-hexane, yielding values between 18 and 40 g/g. A slight decline in the sorption capacity was observed after 10 cycles [44]. As a way to obtain superhydrophobic cellulose-based aerogels, Ma et al. [57] employed a two-step process involving the dissolution of cellulose in a DES and subsequent carbonization of the CNF-prepared aerogels under an argon atmosphere. The resulting aerogels exhibited enhanced surface roughness and hydrophobicity. The sorption capacities of these carbon aerogels for various solvents exhibited values ranging from 65 to 133 g/g, which are higher than those of other reported carbon aerogels. Furthermore, the aerogels demonstrated the ability to be reused for 25 cycles [57].

Peng et al. [59] prepared TEMPO-oxidized cellulose-based materials from the DES-delignified microfibers obtained from corn stalk pith. To make them hydrophobic, they were subjected to a hexamethyldisilazane coating through chemical vapor deposition. The sorbents were tested on various types of oil and organic solvents, demonstrating sorption capacities of 25–34 g/g and a reusability of 20 cycles [59]. Following a similar methodology, Ren et al. [47] used the CNF extracted from corncobs through a DES-based pretreatment to prepare aerogels that were ultimately modified through a chemical vapor deposition process with trimethoxymethylsilane. The hydrophobic aerogels obtained were employed in the absorption of organic solvents (chloroform, n-hexane, 1-propanol, petroleum ether) and oils (soybean oil, olive oil), exhibiting absorption values ranging from 9.6 to 29.8 g/g. In this study, the biodegradability of the aerogels was also confirmed, rendering them highly promising as alternative adsorbents [47]. Thiamngoen et al. [48] employed the aforementioned silanization methodology on DES-delignified fibers derived from cattail. The resulting cellulosic absorbent demonstrated oil absorption capacities of 6–7 g/g [48].

In another study conducted by Wu et al. [46], DES-delignified balsa wood slices were impregnated with carboxymethylcellulose and chitosan quaternary ammonium salt, crosslinked, and subsequently freeze-dried to create a superhydrophilic/underwater superoleophobic wood material. This material was then applied in oil-water separation experiments, achieving a separation efficiency of greater than 99.2%. Moreover, the separation efficiency of kerosene from water remained above 99.5% even after 30 cycles of use [46].

3.1.2 Dye removal

Among the methods for dye removal, adsorption is widely studied for its accessibility, economic viability, and high rate of removal. Cellulose-based materials, due to their abundance and intrinsic characteristics, are extensively studied as economical adsorbents [71].

Yang et al. [63] developed cationic cellulose monoliths through the utilization of a DES for the integration of trimethylammonium groups into the cellulose framework with the objective of enhancing the adsorption capacity for the anionic dye acid red 70. The resulting monolith demonstrated an experimental adsorption capacity of 23.1 mg/g with a total removal of 82.4% and was reused for up to five cycles.

Selkälä et al. [64] introduced aminoguanidine groups into cellulose using DES. This process resulted in the production of cationic CNF, which was then applied to promote the aggregation of tubular halloysite nanotubes adsorbing aqueous anionic dye chrome azurol S. Furthermore, the cationic CNF exhibited synergistic adsorption of the dye, resulting in a removal of 21–24% independently. Notably, the two-step adsorption process demonstrated an efficient removal of 80% of the dye [64]. The same author, in another study conducted in 2020, produced this cationic DES-functionalized cellulose as nanocrystals to create CNC-geopolymer hybrid foams able to adsorb anionic methyl orange. The addition of the functionalized CNC improved the dye removal by up to 84%. The filtrate solution was passed through the CNC-geopolymer columns a total of five times [65].

Wang et al. [50] employed the carboxylated/esterified cellulose-rich residue derived from a DES-pretreatment for the selective adsorption of cationic and anionic dyes. As anticipated, the cationic dye, neutral red, was adsorbed preferentially over the anionic dye, methyl orange, due to the influence of electrostatic interactions. The modified residue demonstrated an adsorption capacity of 300 mg/g and exhibited promising reusability after four cycles [50].

In a recent study, Wen et al. [49] prepared sponges made of cellulose, chitosan, and alginate and tested their adsorption capacity on 10 anionic dyes (methyl blue, eosin Y, eriochrome black T, sunset yellow, brilliant blue, lemon yellow, acid blue 80, amaranth, carmine, and direct red 80) and three cationic dyes (methylene blue, crystal violet, and neutral red). As anticipated for the zeta potential investigation, the fibers demonstrated a positive zeta potential value across most of the examined pH range. This limited their electrostatic attraction with the cationic dyes, thereby reducing their selectivity for these dyes. However, the fibers exhibited selectivity for anionic dyes, with adsorption capacities ranging from 105.9 to 203.7 mg/g, due to electrostatic and hydrogen bond interactions. Additionally, the sponge demonstrated remarkable stability and reusability, retaining 99.2% of its adsorption performance after seven cycles [49].

DES-delignified cellulose-sodium alginate-based microgel microspheres were prepared in a study conducted by Meas et al. [54] for potential use in the adsorption of cationic and anionic dyes. The spheres exhibited adsorption capacities of 160.87 and 141.48 mg/g and adsorption-desorption efficiencies of 96.84% and 95.41% for crystal violet and trypan blue, respectively, after five cycles. The adsorption mechanism may be attributed to the presence of electrostatic interactions, which result in a higher attraction between the cationic dye and the negatively charged adsorbent. However, the adsorption capacity may be linked to pore-filling effects as well [54].

Despite these results, adsorption is not always sufficient for dye removal [72]. The use of cellulose-based materials as both adsorbents and photocatalyst supports has been shown to overcome this limitation and enhance the removal of dyes. Zheng et al. [45] prepared wood-based carriers with a high specific surface area and chemical activity by growing Ag nanoparticles in the pores. Using Ag as a catalyst and $NaBH_4$ as a reducing agent, these wood-based carriers and adsorbents were able to degrade methylene blue with an efficiency of up to 98.3% and a maintained efficiency of up to 89% after seven cycles [45].

In a study conducted by Wang et al. [60], DES-carboxylated cellulose nanofibers were employed as a carbon doping source and biological template for the synthesis of a carbon-nitride-based photocatalyst capable of degrading methyl orange. The results demonstrated that the carbon was capable of almost completely photodegrading the dye within 60 minutes, with the highly condensed lignin in the lignocellulose fibers exhibiting a promoting effect [60].

Zhou et al. [73] developed a novel and reusable composite comprising BiOBr/sulfated lignocellulose nanofibers, which exhibited enhanced visible light adsorption capabilities. This composite was employed for the adsorption and photodegradation of rhodamine B and tetracycline hydrochloride, achieving a 99% degradation of the first organic pollutant within 30 minutes and demonstrating the ability to be used up to five times [68].

3.1.3 Heavy metal ions removal

The release of heavy metal ions into the environment represents a significant environmental concern due to their toxicity and bioaccumulation potential [74]. In this regard, the use of cellulosic-based adsorbents has emerged as a promising area of research, with the potential use of DES-treated cellulose representing a particularly promising path.

In a study conducted by Sirviö and Visanko [66], DES-sulfated lignocellulose nanofibers were used for the adsorption of Pb^{2+} and Cu^{2+} ions. The high content of sulfate groups on the fibers' surface promoted ion exchange between metal ions and the negatively charged sulfate groups. Additionally, carbamate groups acted as chelating agents for metal ions. The maximum experimental adsorption capacity of the sulfated fibers was observed to range between 2.2 and 2.5 mmol/g for Cu^{2+} ions and from 1.1 to 1.5 mmol/g for Pb^{2+} ions [66].

A recently published study by Omer and Fakhre [51] focused on the adsorption of Pb^{2+} ions through DES-delignified and modified microcrystalline cellulose, achieving an adsorption capacity of 50 mg/g. The observed mechanism behind the adsorption of heavy metal ions was found to be likely related to the entrapment of ions within the porous sample, electrostatic attraction, and even the complexation of Pb^{2+} ions with the oxygen atoms of the incorporated functional groups. The partly esterified microcrystalline cellulose demonstrated a removal efficiency of 80.9% after the fourth adsorption-desorption cycle, indicating the potential for scaled-up use [51].

He et al. [55] functionalized DES-delignified beech wood with carboxyl and sulfhydryl groups with the objective of obtaining a three-dimensional wood microfilter with an abundant porous structure and multiple adsorption sites. This was successfully achieved, resulting in a microfilter with the capacity to adsorb Cu^{2+} and Cd^{2+} ions at a notable adsorption capacity of 56.78 and 44.70 mg/g for Cu^{2+} and Cd^{2+}, respectively. Moreover, the cellulose-based microfilter exhibited promising reusability after five cycles [55].

In a recent study, Montenegro et al. [67] prepared both sulfated lignocellulose and cellulose-based hydrogels and sponges for the adsorption of Cu^{2+} ions, comparing the differences in the removal of both nanocellulose-based materials' formats. Similar to Sirviö and Visanko [66], they also reported the influence of carbamate groups in the removal of Cu^{2+} ions through complexation, while sulfate groups promoted the electrostatic interactions between them and the heavy metal ions. The DES-based materials were observed to adsorb Cu^{2+} ions with notable efficacy, exhibiting the sponges a maximum adsorption capacity of 89.75 mg/g and favorable characteristics with regard to ease of handling. However, the sponges demonstrated limited reusability, with only two cycles of reuse as a result of the selected acid desorbing agent, which exerted a physical impact on the structural properties of the materials [67].

3.1.4 Drugs and other pollutants removal

The last few decades have witnessed a notable rise in the prevalence of pharmaceutical pollutants in water bodies, a phenomenon that has attracted considerable attention from researchers [75]. Nanocellulose-based materials have been reported to be very effective for the adsorption of these pollutants [24].

In this regard, Selkälä et al. [61] used a DES to improve the better succinylation of cellulose and then used it as a hydrogel for the removal of salbutamol, a polar, ionizable, and non-volatile micropollutant. The experimental maximum adsorption capacity obtained was 196 mg/g and the adsorption was found to be mainly controlled by the opposite charges of adsorbent and adsorbate [61].

Guo et al. [53] studied the degradation of arbidol, a potential anti-influenza drug recommended for the treatment of COVID-19, via a DES-treated cellulose-based catalytic biochar. This biochar, loaded with a bimetallic oxide, facilitated the catalytic persulfate activation and was able to degrade and remove the drug pollutant with an efficiency of 94.8% in only 60 minutes [53].

The aforementioned BiOBr/DES-sulfated lignocellulose nanofiber composites prepared by Zhou et al. [73] were also used for the adsorption-photocatalytic synergistic removal of tetracycline hydrochloride, achieving 88.9% removal within 60 minutes [73].

Lastly, Yao et al. [62] studied the adsorption of dissolved colloidal substances using polygalacturonic acid as a model pollutant and cationic nanocellulose fibers as adsorbents. The adsorption was achieved by electrostatic interactions between the positive quaternary ammonium groups of the CNF and the negative adsorbate, with a maximum adsorption capacity calculated at 1054 mg/g, a significantly high value compared to other adsorbents for this pollutant [62].

3.2 Air remediation

Cellulose-based materials have emerged as a promising class of biopolymers for the preparation of air filters. These materials offer enhanced diffusion and interception due to their porous structure and a significant number of active functional groups. In addition, they possess biodegradability and reusability properties that make them attractive candidates for use in air filtration applications [23]. Nevertheless, there is still considerable scope for further research in this field, particularly with regard to the application of DES-pretreated and DES-modified cellulose. To date, only a limited number of research articles have been published on this topic with the aforementioned cellulose-based materials.

In a recently published study conducted by Farooq et al. [52], aerogels were prepared using DES-delignified cellulose microfibers obtained from *Firmiana simplex* and polyamideamine-epichlorohydrin resin as a crosslinker. The resulting aerogels were tested for their potential use in filtering PM2.5 and PM10. The filter demonstrated its ability to capture and retain high concentrations of PM, exhibiting excellent purification performance (94.48% and 91.51% for PM10 and PM2.5, respectively) and removing the smoke within 15 minutes of the experiment. Khan et al. [56] assessed the airborne PM retention on nanocellulose-based filters with an induced porous structure. The findings revealed that the maximum filtration efficiency was 32.51% for particles with a diameter of 0.1 μm and 93.63% for particles with a diameter of 1 μm. Besides, an economic assessment was performed for different scenarios to provide insight into the financial viability [56].

DES-sulfated nanocellulose fibers were used in the study conducted by Sun et al. [69] for the synthesis of hybrid aerogels in combination with tourmaline particles. Tourmaline is a hydrated cyclic inorganic silicate permanently self-polarized that interacts with atmospheric oxygen to form NOIs, oxygen molecules with an extra electron known for their air purification activity through sterilization and odor elimination [76]. The aerogels' highly porous structure makes them an appropriate matrix for the particles, enabling their production to 800 cps/cm^3 [69]. In a subsequent study, Sun et al. [70] prepared the same DES-sulfated nanocellulose-based aerogels by combining both tourmaline particles with metal-organic frameworks. This approach not only enabled the production of NOIs with a rate of 700 cps/cm^3, but it also facilitated CO_2 adsorption with an adsorption capacity exceeding 0.9 mmol/g [70].

4. Concluding remarks

The pursuit of sustainable alternatives to replace or minimize the use of harmful substances in the processing of materials is of paramount importance. The synthesis and functionalization of materials by substituting certain particularly harmful chemical agents with more sustainable options such as deep eutectic solvents are fundamental aspects toward reducing their environmental impact. DES have been used in two distinct roles: firstly, as a pretreatment agent due to its delignification capacity and potential for separating cellulose fibers; secondly, as a functionalization agent for its ability to facilitate the incorporation of new functionalities into the cellulose backbone. The underlying objective in both cases has been the production of more efficient, eco-friendly, and sustainable cellulose-based materials that can be utilized in environmental remediation. These materials have demonstrated highly promising results in the treatment of a range of pollutants in diverse media. However, further research is required to fully assess their efficacy in the context of the removal of certain pollutants, such as micropollutants like microplastics in water. Additionally, there is a need for more comprehensive studies on their potential applications in air remediation. Furthermore, the scalability, cost-effectiveness, and environmental impact of both DES and these DES-treated cellulose-based materials remain significant challenges that require further attention to be considered sustainable alternatives.

Acknowledgements

The authors acknowledge the financial support provided by Ministerio de Ciencia, Innovación y Universidades (MCIN/AEI/10.13039/501100011033) from Spain through TED2021-129920A-C42 and PID2023-152932OB-I00 projects.

Author details

Remedios Montenegro*, Zoilo González and Alejandro Rodríguez Pascual
BioPrEn Group (RNM940), Chemical Engineering Department, Instituto Químico para la Energía y el Medioambiente (IQUEMA), Faculty of Science, Universidad de Córdoba, Córdoba, Spain

*Address all correspondence to: z12moelr@uco.es

References

[1] Landrigan PJ, Fuller R, Acosta NJR, Adeyi O, Arnold R, Basu N(N), et al. The lancet commission on pollution and health. Lancet. 2018;**391**(10119):462-512. DOI: 10.1016/S0140-6736(17)32345-0

[2] Gaur N, Sharma S, Yadav N. Environmental pollution. In: Green Chemistry Approaches to Environmental Sustainability: Status, Challenges and Prospective. Amsterdam: Elsevier; 2023. pp. 23-41. DOI: 10.1016/B978-0-443-18959-3.00010-0

[3] Yadav P, Usha K, Singh B. Air pollution mitigation and global dimming: A challenge to agriculture under changing climate. In: Climate Change and Crop Stress: Molecules to Ecosystems. Amsterdam: Elsevier; 2022:271-298. DOI: 10.1016/B978-0-12-816091-6.00015-8

[4] Health Effects Institute. State of Global Air 2024: A Special Report on Global Exposure to Air Pollution and Its Health Impacts, With a Focus on Children's Health. Boston, MA: Health Effects Institute; 2024. Available from: https://www.stateofglobalair.org/resources/report/state-global-air-report-2024

[5] Goswami R, Singh S, Narasimhappa P, Ramamurthy PC, Mishra A, Mishra PK, et al. Nanocellulose: A comprehensive review investigating its potential as an innovative material for water remediation. International Journal of Biological Macromolecules. 2024;**254**:127465. DOI: 10.1016/J.IJBIOMAC.2023.127465

[6] Fuller R, Landrigan PJ, Balakrishnan K, Bathan G, Bose-O'Reilly S, Brauer M, et al. Pollution and health: A progress update. Lancet Planet Health. 2022;**6**(6):e535-e547. DOI: 10.1016/S2542-5196(22)00090-0

[7] Münzel T, Hahad O, Daiber A, Landrigan PJ. Soil and water pollution and human health: What should cardiologists worry about? Cardiovascular Research. 2022;**119**(2):440. DOI: 10.1093/CVR/CVAC082

[8] Wu P-Y, Lasek J, Głód K, Supernok K, Wu JC-S. Photo-catalytic and thermal-assisted air purification of VOC and NOx simultaneously by pilot-scale photo-reactor. Applied Catalysis O: Open. 2024a;**188**:206945. DOI: 10.1016/J.APCATO.2024.206945

[9] Ren H, Koshy P, Chen WF, Qi S, Sorrell CC. Photocatalytic materials and technologies for air purification. Journal of Hazardous Materials. 2017;**325**:340-366. DOI: 10.1016/J.JHAZMAT.2016.08.072

[10] Li J, Gao H, Lan C, Nie L, Liu D, Lu X, et al. Plasma air filtration system for intercepting and inactivation of pathogenic microbial aerosols. Journal of Environmental Chemical Engineering. 2023a;**11**(5):110728. DOI: 10.1016/J.JECE.2023.110728

[11] Yan S, Liu C, Hou L, Wang B, Zhang Y. A new filterless indoor air purifier for particulate matter and bioaerosol based on heterogeneous condensation. Environmental Research. 2023;**218**:115034. DOI: 10.1016/J.ENVRES.2022.115034

[12] Zhang Y, Han Y, Ji X, Zang D, Qiao L, Sheng Z, et al. Continuous air purification by aqueous interface filtration and absorption. Nature. 2022;**610**(7930):74-80. DOI: 10.1038/s41586-022-05124-y

[13] Law CK, Lai JHK, Ma XD, Sze-To GN. Enhancing indoor air quality: Examination of formaldehyde adsorption efficiency of portable air cleaner fitted with chemically-treated activated carbon filters. Building and Environment. 2024;**263**:111823. DOI: 10.1016/J.BUILDENV.2024.111823

[14] Nguyen-Tri P, Nguyen TA, Nguyen TV. Nanomaterial for air remediation: An introduction. Nanomaterials for Air Remediation. 2020;**1**:3-8. DOI: 10.1016/B978-0-12-818821-7.00001-4

[15] De la Torre-Bayo JJ, Martín-Pascual J, Torres-Rojo JC, Zamorano M. Characterization of screenings from urban wastewater treatment plants: Alternative approaches to landfill disposal. Journal of Cleaner Production. 2022;**380**:134884. DOI: 10.1016/J.JCLEPRO.2022.134884

[16] Jiang JQ. The role of coagulation in water treatment. Current Opinion in Chemical Engineering. 2015;**8**:36-44. DOI: 10.1016/J.COCHE.2015.01.008

[17] Saravanathamizhan R, Perarasu VT. Improvement of biodegradability index of industrial wastewater using different pretreatment techniques. In: Wastewater Treatment: Cutting Edge Molecular Tools, Techniques and Applied Aspects. Amsterdam: Elsevier; 2021:103-136. DOI: 10.1016/B978-0-12-821881-5.00006-4

[18] Saravanan A, Deivayanai VC, Kumar PS, Rangasamy G, Hemavathy RV, Harshana T, et al. A detailed review on advanced oxidation process in treatment of wastewater: Mechanism, challenges and future outlook. Chemosphere. 2022;**308**:136524. DOI: 10.1016/J.CHEMOSPHERE.2022.136524

[19] Kamar MT, Elattar H, Mahmoud AS, Peters RW, Mostafa MK. A critical review of state-of-the-art technologies for electroplating wastewater treatment. International Journal of Environmental Analytical Chemistry. 2024;**104**(16):4143-4176. DOI: 10.1080/03067319.2022.2098486

[20] Yaashikaa PR, Senthil Kumar P, Karishma S. Review on biopolymers and composites – Evolving material as adsorbents in removal of environmental pollutants. Environmental Research. 2022;**212**:113114. DOI: 10.1016/J.ENVRES.2022.113114

[21] Jain R, Camarillo MK, Stringfellow WT. Threats. In: Drinking Water Security for Engineers, Planners, and Managers. Oxford: Butterworth-Heinemann; 2014. pp. 45-67. DOI: 10.1016/B978-0-12-411466-1.00003-3

[22] Liu Y, Wang H, Cui Y, Chen N. Removal of copper ions from wastewater: A review. International Journal of Environmental Research and Public Health. 2023;**20**(5):3885. DOI: 10.3390/IJERPH20053885

[23] Rana AK, Mostafavi E, Alsanie WF, Siwal SS, Thakur VK. Cellulose-based materials for air purification: A review. Industrial Crops and Products. 2023;**194**:116331. DOI: 10.1016/J.INDCROP.2023.116331

[24] Abdelhamid HN, Mathew AP. Cellulose-based materials for water remediation: Adsorption, catalysis, and antifouling. Frontiers in Chemical Engineering. 2021;**3**:790314. DOI: 10.3389/FCENG.2021.790314

[25] Xiao W, Sun R, Hu S, Meng C, Xie B, Yi M, et al. Recent advances and future perspective on lignocellulose-based materials as adsorbents in diverse water

treatment applications. International Journal of Biological Macromolecules. 2023;**253**:126984. DOI: 10.1016/J.IJBIOMAC.2023.126984

[26] Shahzad A, Ullah MW, Ali J, Aziz K, Javed MA, Shi Z, et al. The versatility of nanocellulose, modification strategies, and its current progress in wastewater treatment and environmental remediation. Science of The Total Environment. 2023;**858**:159937. DOI: 10.1016/J.SCITOTENV.2022.159937

[27] Mai T, Wang PL, Ma MG. Promising cellulose-based aerogel composites: Preparation methods and advanced applications. Composites Part B: Engineering. 2024;**281**:111559. DOI: 10.1016/J.COMPOSITESB.2024.111559

[28] Ghasemlou M, Daver F, Ivanova EP, Habibi Y, Adhikari B. Surface modifications of nanocellulose: From synthesis to high-performance nanocomposites. Progress in Polymer Science. 2021;**119**:101418. DOI: 10.1016/J.PROGPOLYMSCI.2021.101418

[29] Voisin H, Bergström L, Liu P, Mathew AP. Nanocellulose-based materials for water purification. Nanomaterials. 2017;**7**(3):57. DOI: 10.3390/NANO7030057

[30] Mahfoudhi N, Boufi S. Nanocellulose as a novel nanostructured adsorbent for environmental remediation: A review. Cellulose. 2017;**24**(3):1171-1197. DOI: 10.1007/S10570-017-1194-0

[31] Faiz Norrrahim MN, Mohd Kasim NA, Knight VF, Mohamad Misenan MS, Janudin N, Ahmad Shah NA, et al. Nanocellulose: A bioadsorbent for chemical contaminant remediation. RSC Advances. 2021;**11**(13):7347-7368. DOI: 10.1039/D0RA08005E

[32] Sharma A, Anjana RH, Goswami S. A comprehensive review on the heavy metal removal for water remediation by the application of lignocellulosic biomass-derived nanocellulose. Journal of Polymers and the Environment. 2022;**30**(1):1-18. DOI: 10.1007/S10924-021-02185-4

[33] Sjahro N, Yunus R, Abdullah LC, Rashid SA, Asis AJ, Akhlisah ZN. Recent advances in the application of cellulose derivatives for removal of contaminants from aquatic environments. Cellulose. 2021;**28**(12):7521-7557. DOI: 10.1007/S10570-021-03985-6

[34] Zhang L, Guo L, Wei G. Recent advances in the fabrication and environmental science applications of cellulose nanofibril-based functional materials. Maternité. 2021;**14**(18):5390. DOI: 10.3390/MA14185390

[35] Hamidon TS, Adnan R, Haafiz MKM, Hussin MH. Cellulose-based beads for the adsorptive removal of wastewater effluents: A review. Environmental Chemistry Letters. 2022;**20**(3):1965-2017. DOI: 10.1007/S10311-022-01401-4

[36] Martins MAR, Pinho SP, Coutinho JAP. Insights into the nature of eutectic and deep eutectic mixtures. Journal of Solution Chemistry. 2019;**48**(7):962-982. DOI: 10.1007/S10953-018-0793-1

[37] Zaib Q, Masoumi Z, Aich N, Kyung D. Review of the synthesis and applications of deep eutectic solvent-functionalized adsorbents for water treatment. Journal of Environmental Chemical Engineering. 2023;**11**(3):110214. DOI: 10.1016/J.JECE.2023.110214

[38] Hansen BB, Spittle S, Chen B, Poe D, Zhang Y, Klein JM, et al. Deep eutectic solvents: A review of fundamentals

and applications. Chemical Reviews. 2021;**121**(3):1232-1285. DOI: 10.1021/ACS.CHEMREV.0C00385

[39] Almeida RO, Maloney TC, Gamelas JAF. Production of functionalized nanocelluloses from different sources using deep eutectic solvents and their applications. Industrial Crops and Products. 2023;**199**:116583. DOI: 10.1016/J.INDCROP.2023.11658

[40] Elhamarnah Y, Qiblawey H, Nasser M. A review on deep eutectic solvents as the emerging class of green solvents for membrane fabrication and separations. Journal of Molecular Liquids. 2024;**398**:124250. DOI: 10.1016/J.MOLLIQ.2024.124250

[41] Contreras-Gámez M del M, Galán-Martín Á, Seixas N, da Costa Lopes AM, Silvestre A, Castro E. Deep eutectic solvents for improved biomass pretreatment: Current status and future prospective towards sustainable processes. Bioresource Technology. 2023;**369**:128396. DOI: 10.1016/J.BIORTECH.2022.128396

[42] Li P, Yang C, Jiang Z, Jin Y, Wu W. Lignocellulose pretreatment by deep eutectic solvents and related technologies: A review. Journal of Bioresources and Bioproducts. 2023b;**8**(1):33-44. DOI: 10.1016/J.JOBAB.2022.11.004

[43] Aziz T, Farid A, Haq F, Kiran M, Ullah A, Zhang K, et al. A review on the modification of cellulose and its applications. Polymers. 2022;**14**(15):3206. DOI: 10.3390/POLYM14153206

[44] Yang R, Cao Q, Liang Y, Hong S, Xia C, Wu Y, et al. High capacity oil absorbent wood prepared through eco-friendly deep eutectic solvent delignification. Chemical Engineering Journal. 2020;**401**:126150. DOI: 10.1016/J.CEJ.2020.126150

[45] Zheng G, Kong F, Fu P, Li X. Preparation of modified wood-based metal nanoparticles synthesized in situ and efficient removal of organic pollutants from wastewater. Colloids Surfaces A: Physicochemical and Engineering Aspects. 2024;**695**:134246. DOI: 10.1016/J.COLSURFA.2024.134246

[46] Wu J, Li H, Wang P, Zhang Y, Tan M, Qin W, et al. Natural polyelectrolyte-functionalized superhydrophilic and antibacterial wood for efficient oil/water separation. Journal of Porous Materials. 2024b;**31**(2):709-725. DOI: 10.1007/S10934-023-01548-7

[47] Ren M, Ji X, Kong F, Zhou C, Yagoub AEGA, Liang J, et al. Preparation and characterization of bio-inspired full-biomass-derived aerogel with vertically aligned structure. Chemical Engineering Journal. 2023;**475**:146425. DOI: 10.1016/J.CEJ.2023.146425

[48] Thiamngoen P, Pinpatthanapong K, Juntadech NC, Rangseesuriyachai T. Preparation of cellulose fiber from cattail using two-stage pretreatment and choline chloride-oxalic acid deep eutectic solvent extraction for oil adsorption. Environmental Processes. 2024;**11**(3):1-17. DOI: 10.1007/S40710-024-00702-W

[49] Wen Z, Gao D, Niu H, Lin J, Li Z, Zeng J, et al. Three-dimensional porous adsorbent based on chitosan-alginate-cellulose sponge for selective and efficient removal of anionic dyes. Journal of Environmental Chemical Engineering. 2023;**11**(5):110831. DOI: 10.1016/J.JECE.2023.110831

[50] Wang S, Liu M, Ge W, Jin C, Bi W. Sustainable and efficient extraction of lignin from wood meal using a deep eutectic system and adsorption

of neutral red dye with the extraction residue. Journal of Cleaner Production. 2023;**430**:139687. DOI: 10.1016/J.JCLEPRO.2023.139687

[51] Omer SK, Fakhre NA. Removal of aqueous Pb^{2+} on partially esterified cellulose obtained by almond shell dissolution using deep eutectic solvent. Iran Journal of Science. 2023;**47**(3):749-764. DOI: 10.1007/S40995-023-01419-5

[52] Farooq A, Yang H, Ding Z, Bu F, Guo M, Sun W, et al. Exploring the versatility of biodegradable biomass aerogels: In-depth evaluation of Firmiana simplex bark microfibers depolymerized by deep eutectic solvent. International Journal of Biological Macromolecules. 2024;**275**:133629. DOI: 10.1016/J.IJBIOMAC.2024.133629

[53] Guo J, Xu J, Liu X, Dai L, Zhang C, Xiao X, et al. Enabling dual valorization of lignocellulose by fluorescent lignin carbon dots and biochar-supported persulfate activation: Towards waste-treats-pollutant. Journal of Hazardous Materials. 2022;**435**:129072. DOI: 10.1016/J.JHAZMAT.2022.129072

[54] Meas A, Wi E, Chang M, Hwang HS. Carboxylmethyl cellulose produced from wood sawdust for improving properties of sodium alginate hydrogel in dye adsorption. Separation and Purification Technology. 2024;**341**:126906. DOI: 10.1016/J.SEPPUR.2024.126906

[55] He LT, Zhao LD, Sun W, Fang J, Liu XW, Qi JJ, et al. 3D wood microfilter for fast and efficient removal of heavy metal ions from water. Langmuir. 2023;**39**(43):15319-15327. DOI: 10.1021/ACS.LANGMUIR.3C02129

[56] Khan MJ, Chaipanya RJ, Suksomboon S, Sonyeam J, Posoknistakul P, Charnnok B, et al. 3D porous nanocellulose based filter from palm bunch using tert-butyl alcohol-assisted pore inducive technique for airborne particulate matter retention. ChemNanoMat. 2024a;**10**(11):e202400129. DOI: 10.1002/CNMA.202400129

[57] Ma X, Zhou S, Li J, Xie F, Yang H, Wang C, et al. Natural microfibrils/regenerated cellulose-based carbon aerogel for highly efficient oil/water separation. Journal of Hazardous Materials. 2023;**454**:131397. DOI: 10.1016/J.JHAZMAT.2023.131397

[58] Laitinen O, Suopajärvi T, Österberg M, Liimatainen H. Hydrophobic, superabsorbing aerogels from choline chloride-based deep eutectic solvent pretreated and silylated cellulose nanofibrils for selective oil removal. ACS Applied Materials & Interfaces. 2017;**9**(29):25029-25037. DOI: 10.1021/ACSAMI.7B06304

[59] Peng D, Zhao J, Liang X, Guo X, Li H. Corn stalk pith-based hydrophobic aerogel for efficient oil sorption. Journal of Hazardous Materials. 2023;**448**:130954. DOI: 10.1016/J.JHAZMAT.2023.130954

[60] Wang B, Wang J, Hu Z, Zhu A-L, Shen X, Cao X, et al. Harnessing renewable lignocellulosic potential for sustainable wastewater purification. Research. 2024;7:0347. DOI: 10.34133/RESEARCH.0347

[61] Selkälä T, Suopajärvi T, Sirviö JA, Luukkonen T, Lorite GS, Kalliola S, et al. Rapid uptake of pharmaceutical salbutamol from aqueous solutions with anionic cellulose nanofibrils: The importance of pH and colloidal stability in the interaction with ionizable pollutants. Chemical Engineering Journal. 2018;**350**:378-385. DOI: 10.1016/j.cej.2018.05.163

[62] Yao L, Zou X, Zhou S, Zhu H, Chen G, Wang S, et al. Cationic lignocellulose nanofibers from agricultural waste as high-performing adsorbents for the removal of dissolved and colloidal substances. Polymers (Basel). 2022;**14**(5):910. DOI: 10.3390/POLYM14050910

[63] Yang Z, Asoh TA, Uyama H. Cationic functionalization of cellulose monoliths using a urea-choline based deep eutectic solvent and their applications. Polymer Degradation and Stability. 2019;**160**:126-135. DOI: 10.1016/J.POLYMDEGRADSTAB. 2018.12.015

[64] Selkälä T, Suopajärvi T, Sirviö JA, Luukkonen T, Kinnunen P, Kling KI, et al. Efficient entrapment and separation of anionic pollutants from aqueous solutions by sequential combination of cellulose nanofibrils and halloysite nanotubes. Chemical Engineering Journal. 2019;**374**:1013-1024. DOI: 10.1016/J.CEJ.2019.06.008

[65] Selkälä T, Suopajärvi T, Sirviö JA, Luukkonen T, Kinnunen P, De Carvalho ALCB, et al. Surface modification of cured inorganic foams with cationic cellulose nanocrystals and their use as reactive filter media for anionic dye removal. ACS Applied Materials & Interfaces. 2020;**12**(24):27745-27757. DOI: 10.1021/ACSAMI.0C05927

[66] Sirviö JA, Visanko M. Lignin-rich sulfated wood nanofibers as high-performing adsorbents for the removal of lead and copper from water. Journal of Hazardous Materials. 2020;**383**:121174. DOI: 10.1016/J.JHAZMAT.2019.121174

[67] Montenegro R, Rincón E, Rodríguez A, González Z. Manufacturing sulfated cellulose nanofibers using a unique combined DES-based pretreatment-functionalization protocol for metal ion decontamination through porous adsorbents. Carbohydrate Polymers. 2025;**349**:122974. DOI: 10.1016/J.CARBPOL.2024.122974

[68] Zou X, Yao L, Zhou S, Chen G, Wang S, Liu X, et al. Sulfated lignocellulose nanofibril based composite aerogel towards adsorption-photocatalytic removal of tetracycline. Carbohydrate Polymers. 2022;**296**:119970. DOI: 10.1016/j.carbpol.2022.119970

[69] Sun X, Yu Q, Wang F, Hu S, Zhou J, Liu Y, et al. Sustainable lignocellulose aerogel for air purifier with thermal insulation, flame retardancy, mechanical strength, and its life cycle assessment. International Journal of Biological Macromolecules. 2024a;**257**:128599. DOI: 10.1016/J.IJBIOMAC.2023.128599

[70] Sun X, Yu Q, Wang F, Sun M, Hu S, Zhu J, et al. Eco-friendly tourmaline@MOF lignocellulose aerogel with favorable fire retardancy and smoke suppression for insulation materials. ACS Sustainable Chemistry & Engineering. 2024b;**12**(28):10517-10527. DOI: 10.1021/ACSSUSCHEMENG.4C02904

[71] Kausar A, Zohra ST, Ijaz S, Iqbal M, Iqbal J, Bibi I, et al. Cellulose-based materials and their adsorptive removal efficiency for dyes: A review. International Journal of Biological Macromolecules. 2023;**224**:1337-1355. DOI: 10.1016/J.IJBIOMAC.2022.10.220

[72] Khan S, Noor T, Iqbal N, Yaqoob L. Photocatalytic dye degradation from textile wastewater: A review. ACS Omega. 2024b;**9**(20):21751-21767. DOI: 10.1021/ACSOMEGA.4C00887

[73] Zhou X, Liu Y, Liu X, Xu M, Shi J, Huang Q. Sulfated lignocellulose nanofibril supported flower-like BiOBr with oxygen vacancies towards

absorption–photocatalytic synergetic removal of organic pollutants. New Journal of Chemistry. 2024;**48**(9):3910-3923. DOI: 10.1039/D3NJ05486A

[74] Qasem NAA, Mohammed RH, Lawal DU. Removal of heavy metal ions from wastewater: A comprehensive and critical review. Clean Water. 2021;**4**(1):1-15. DOI: 10.1038/s41545-021-00127-0

[75] Sirés I, Brillas E. Remediation of water pollution caused by pharmaceutical residues based on electrochemical separation and degradation technologies: A review. Environment International. 2012;**40**(1):212-229. DOI: 10.1016/J.ENVINT.2011.07.012

[76] Gao L, Qiu Z, Gan W, Zhan X, Li J, Qiang T. Negative oxygen ions production by superamphiphobic and antibacterial TiO_2/Cu_2O composite film anchored on wooden substrates. Scientific Reports. 2016;**6**(1):1-10. DOI: 10.1038/srep26055

Chapter 6

Biobased Composites: A Sustainable Pathway for Modern Applications

Behrooz Maleki, Pouya Ghamari Kargar and Samaneh Sedigh Ashrafi

Abstract

Biobased composites, which combine natural fibers with biopolymers, represent a sustainable alternative to traditional petroleum-based materials. These composites leverage renewable resources, reducing environmental impact and enhancing biodegradability. The integration of agricultural waste, such as hemp, flax, and jute, with biopolymers such as polylactic acid (PLA) and polyhydroxyalkanoates (PHA) fosters advancements in material performance, including improved strength, stiffness, and thermal properties. Applications span various industries, including automotive, construction, and consumer goods, where they contribute to lightweight structures and energy-efficient designs. Furthermore, the use of biobased composites supports circular economy principles by minimizing waste and promoting resource efficiency. Research continues to focus on optimizing processing techniques and enhancing properties through modifications in fiber treatment and matrix formulation. As the demand for eco-friendly materials grows, biobased composites offer a promising pathway toward sustainable development, aligning with global efforts to reduce carbon footprints and promote environmental stewardship. However, the current scientific knowledge regarding the long-term durability performance of biocomposites in the open access literature is very limited, which can be further investigated. Biocomposites can be developed using a variety of natural fibers, such as hemp, flax, and jute, which are integrated with biopolymers such as polylactic acid (PLA) and polyhydroxyalkanoates (PHA). For instance, in the automotive sector, hemp fibers have been used to reinforce biopolymer matrices, resulting in lightweight components that enhance fuel efficiency. In construction, flax fibers have been incorporated into cementitious materials, improving their mechanical properties while maintaining sustainability. These advancements demonstrate significant improvements over traditional materials, including enhanced strength, reduced weight, and increased biodegradability.

Keywords: biobased composites, natural fibers, cellulose, biopolymers, sustainability, renewable resources, circular economy

1. Introduction

Global production of plastics has surged, with an estimated 450 million metric tons manufactured annually [1]. Plastics, characterized by their polymeric structures, offer a unique combination of advantageous properties such as high mechanical strength, chemical inertness, low density, and affordability. These attributes make them highly desirable for various applications, particularly single-use products. However, the widespread use of these materials has led to significant environmental challenges, prompting global concern about their impact [2]. The inert nature of commonly used polymers renders them nonbiodegradable, leading to limited and ineffective end-of-life (EoL) solutions. This issue is further compounded by rising raw material costs, which have added financial strain to the plastic industry [3]. Conventional synthetic polymers exacerbate environmental problems due to their prolonged accumulation in ecosystems during the EoL phase [4].

In response to the depletion of petroleum resources and increasing ecological awareness, there is a growing push to develop new materials that balance sustainability with reduced reliance on fossil fuels [5]. This transition is supported by rising consumer demand for eco-friendly products, driven by green marketing initiatives that promote recycling, social responsibility, and shifting consumer preferences toward sustainable alternatives [6]. To address these challenges, there is an urgent need for sustainable technologies to mitigate the environmental footprint of polymeric materials [7]. Biopolymers, derived from renewable resources, provide a promising solution. These materials are environmentally friendly and typically biodegradable, making them suitable substitutes for petroleum-based plastics [8]. However, their inherent limitations—such as lower impact resistance, tensile strength, permeability, and thermal stability—necessitate further enhancement to match the performance of conventional materials [9]. A viable strategy for improving biopolymers involves reinforcing them with micro- or nanoscale additives, resulting in advanced biopolymer composites with the enhanced properties [10]. These eco-friendly composites exhibit superior mechanical, thermal, and barrier properties, making them ideal for a range of applications, including automotive, aerospace, biomedical, and food packaging sectors [11–14]. Compared to unmodified polymers, these composites are lighter and demonstrate better performance in terms of rheology, durability, and biodegradability [15].

Abdillah and Charles [16] investigated the development of arrowroot starch/iota-carrageenan films at various concentrations. Their findings demonstrated that these composites outperformed pristine cassava starch films, exhibiting superior mechanical strength, improved barrier properties, tensile performance, and reduced swelling characteristics [16]. Similarly, Khoo, Chow, and Ismail [17] conducted a comparative study on the properties of cellulose nanocrystals derived from sugarcane bagasse fibers (SBFCNC) and microcrystalline cellulose (MCC-CNC) when incorporated into poly(lactic acid) (PLA) matrices. Their research highlighted that the type of cellulose nanocrystal significantly influenced the tensile strength, thermal stability, and UV protection capabilities of the PLA nanocomposites. Among the two, SBFCNC displayed superior performance in tensile properties and UV shielding, emphasizing its potential as an effective bio-based nanofiller for PLA applications [17]. Advancements in this field have showcased the increasing utilization of biodegradable polymer composites due to their enhanced mechanical properties, compatibility with diverse matrices, and eco-friendly biodegradability [18, 19].

This chapter aims to provide a comprehensive overview of biopolymers, focusing on their fundamental properties and the integration of environmentally sustainable micro- and nano-reinforcements. It also explores the chemical and physical modifications of reinforcements to improve performance, alongside traditional and innovative processing techniques for fabricating biocomposites. Furthermore, the chapter discusses advanced applications of these biocomposites in addressing the growing demand for sustainable materials in various industries. The production of biocomposites involves various methods, including melt blending, extrusion, and injection molding, which allow for the effective integration of natural fibers into polymer matrices.

Melt blending involves mixing natural fibers with molten polymer, ensuring uniform distribution and strong interfacial adhesion. Extrusion, on the other hand, forces the polymer and natural fibers through a die to create continuous profiles, enhancing the material's consistency and mechanical properties. Injection molding injects the molten polymer-fiber mixture into molds to produce complex shapes with high precision. These techniques not only enhance the mechanical properties of the resulting materials but also ensure that the biocomposites are suitable for large-scale production.

2. Polymers and their properties

A polymer is a high-molecular-weight macromolecule composed of repeating structural units, known as monomers, which are typically linked by covalent bonds. Polymers can be broadly categorized into three types: synthetic, semisynthetic, and natural. Synthetic polymers are produced in laboratories through polymerization, a process in which monomers react under controlled conditions to form polymer chains. Common examples of synthetic polymers include nylon, polyethylene, polystyrene, synthetic rubber, polyvinyl chloride, and Teflon. In contrast, natural polymers occur naturally, sourced primarily from plants or animals. When natural polymers undergo chemical modifications to alter their properties, they are referred to as semisynthetic polymers [20]. In recent years, industries across various sectors have prioritized reducing reliance on fossil fuel-derived materials, driving the demand for sustainable and environmentally friendly alternatives [21]. The majority of conventional polymers are derived from fossil fuels and are nonbiodegradable, creating significant environmental challenges. In response, natural polymers have garnered increasing attention due to their abundance, renewability, and ease of access. Compared to synthetic and semisynthetic counterparts, natural polymers offer several advantages, including lower costs, biocompatibility, the potential for chemical modifications, and inherent biodegradability. These qualities make natural polymers highly suitable for a wide array of applications. For instance, they are extensively used in medicine [22], pharmacology [20], and industries such as packaging [18], cosmetics [23], electronics [24], agriculture [25], water treatment [26], and biosensor development [27].

The term "biopolymer" is commonly used to describe natural polymers, though it is occasionally misapplied. Terms like biopolymers, bio-based polymers, bioplastics, and biodegradable polymers are often used interchangeably, but they can be ambiguous and lead to confusion. Biopolymers typically refer to polymers derived from natural sources and are usually biodegradable. However, this definition does not extend to synthesized biodegradable polymers. In contrast, biodegradable polymers are characterized by their ability to break down over time, regardless of whether they are naturally occurring or synthetically produced. Polymers that are biodegradable or nonbiodegradable but are derived from renewable resources are considered bio-based polymers. Lastly, bioplastics, while often biodegradable, do not necessarily originate from biological sources; for example, polycaprolactone (PCL) and polybutylene succinate (PBS) are petroleum-based polymers that are biodegradable [22]. Biopolymers can be divided into two main categories: those derived from animals and those from plants. Both types are produced through enzymatic processes, which are vital for their metabolism. Naturally occurring biopolymers include polysaccharides, proteins, lipids, microbial polyesters, polyphenols, and essential oils [22]. Among these, polysaccharides are the most abundant macromolecules in the biosphere and play a key role in the structural integrity of both plants and animals [28]. They are widely available in nature from sources like algae, plants, and animals and can also be synthesized using recombinant DNA technology [20]. Examples of animal-derived polysaccharides include chitin, alginates, carrageenan, psyllium, and xanthan gum, while plant-based polysaccharides include hemicellulose, cellulose, pectin, glucomannan, agar, starch, and inulin [29].

The most significant biopolymers in plants are the key components of natural plant fibers, which are found in the cell walls of plants. These biopolymers provide the mechanical properties necessary for the plant's structural functions and activities [30]. In many plant fibers, the specific strength and stiffness are comparable to those of synthetic fibers. Additionally, when comparing densities, synthetic fibers tend to have higher densities (approximately 2.5 g/cm^3), while plant fibers typically have lower densities (around 1.4 g/cm^3) [30, 31]. Given that weight is a critical factor for structural applications, natural plant fibers are highly valuable for reinforcing composite materials, as they offer high strength and low density. Furthermore, natural plant fibers are environmentally friendly due to their CO_2 neutrality, biodegradability, renewability, and recyclability. These fibers stand in stark contrast to synthetic and petroleum-based fibers, which contribute to environmental pollution and are nonbiodegradable [32].

2.1 Composite materials

Composite materials are defined as combinations of two or more materials to create a final product with superior properties compared to each individual component. These materials consist of a matrix, which serves as the continuous phase, and reinforcement elements, which form the discontinuous phase. The reinforcements can be polymeric, metallic, or ceramic. The matrix and reinforcement phases work together to enhance the overall material's performance. The combinations of materials used to produce composites are nearly limitless, offering the flexibility to create various formulations. The properties of these composites depend on factors such as the nature of the matrix, types of reinforcements, their relative proportions, and the processing techniques employed [33–35]. The most widely used composites are large diffusion composites, which account for more than 95% of current production. These materials are primarily utilized in the transportation industry, particularly in automobiles, as well as in electrical construction. Typically, these composites are made from fiberglass combined with unsaturated or phenolic polyester resins, which are nonbiodegradable synthetic materials that pose significant environmental concerns [33]. In the twenty-first century, one of the main challenges researchers' faces is finding ways to minimize the environmental impact of these hazardous materials [36]. As a response to this issue, natural fibers are increasingly being considered as an alternative to traditional synthetic fibers in the production of composite materials [37].

Research has demonstrated that natural fibers, with mechanical properties comparable to synthetic fibers—such as high strength, stiffness, and low density—are suitable for producing hybrid natural fiber composites. These composites can replace synthetic fiber-reinforced materials in structural and semistructural applications [38]. Furthermore, natural fibers are environmentally friendly due to their biodegradability, renewability, and recyclability [31]. As a result, these composites are gaining traction across various industries, including automotive, construction, aerospace, medical, and electronics sectors [21, 37]. The demand for natural fiber-reinforced composites to reduce weight in transportation is growing, particularly within the automotive and aerospace industries [21]. Ramasubbu and Madasamy investigated polymer composites reinforced with natural fibers for automotive applications. They compared the mechanical properties and water absorption characteristics of three composites: a hybrid composite (HC) with sisal and kenaf fibers in an epoxy matrix, a sisal fiber composite (SFC), and a kenaf fiber composite (KFC). Their findings showed that the HC exhibited superior properties compared to the other composites, making it suitable for automotive applications without significant weight penalties [39].

In another study, Sarikaya, Çallioğlu, and Demirel examined the impact of different natural fibers as reinforcements in composite materials with an epoxy resin matrix. The composites incorporated birch, palm, and eucalyptus fibers. The study revealed that the type of fiber used greatly influenced the mechanical properties, emphasizing the importance of fiber selection in composite performance [21]. Otto et al. [40] also studied the effect of various natural fibers on the development of polyurethane (PU)-based composites. They replaced up to 20% of polyethylene glycol with a mix of three natural fibers-sugarcane bagasse, sisal, and rice husk. The mechanical properties of these composites showed that sugarcane, rice husk, and their hybrid composites outperformed sisal-only composites. These natural fiber composites demonstrated potential for use in shock-absorbing and padding applications, providing an economical and sustainable alternative to conventional PU composites [40]. Ashik, Sharma, and Gupta compared the performance of composite materials with an epoxy resin matrix reinforced with two types of fibers: coconut fiber and fiberglass. Mechanical testing and water absorption assessments revealed that composites reinforced with a combination of coconut and glass fibers exhibited increased water absorption as the proportion of coconut fiber increased. Additionally, the composites demonstrated higher flexural strength. These findings suggest that incorporating coconut fiber alongside glass fiber not only enhances the strength of the composite but also offers a viable alternative to using fiberglass alone [41].

Natural fibers such as jute, sisal, and bamboo are increasingly being used as reinforcements in biocomposites due to their high strength-to-weight ratio and biodegradability. For example, jute fibers exhibit excellent tensile strength and stiffness, making them suitable for applications in packaging and construction. Compared to synthetic fibers, natural fibers offer a more sustainable alternative, with lower environmental impact and enhanced recyclability. The production of biocomposites involves various methods, including melt blending, extrusion, and injection molding, which allow for the effective integration of natural fibers into polymer matrices. Melt blending involves mixing natural fibers with molten polymer, ensuring uniform distribution and strong interfacial adhesion. Extrusion, on the other hand, forces the polymer and natural fibers through a die to create continuous profiles, enhancing the material's consistency and mechanical properties. Injection molding injects the molten polymer-fiber mixture into molds to produce complex shapes with high precision. These techniques not only enhance the mechanical properties of the resulting materials but also ensure that the biocomposites are suitable for large-scale production [42].

2.1.1 Natural fibers in biocomposites

Natural fibers such as jute, sisal, and bamboo are increasingly being used as reinforcements in biocomposites due to their high strength-to-weight ratio and biodegradability. These fibers are derived from various plant sources and offer a range of mechanical properties that can be tailored to specific applications.

2.1.2 Jute fibers

Jute fibers are known for their excellent tensile strength and stiffness. They are commonly used in applications such as packaging and construction, where their high strength-to-weight ratio provides significant advantages. Jute fibers also exhibit good thermal stability and are biodegradable, making them an environmentally friendly option.

2.1.3 Sisal fibers

Sisal fibers, derived from the leaves of the *Agave sisalana* plant, are another popular choice for biocomposites. They offer high tensile strength and durability, making them suitable for use in automotive components and other high-performance applications. Sisal fibers are also resistant to moisture and have good dimensional stability.

2.1.4 Bamboo fibers

Bamboo fibers are known for their exceptional strength and flexibility. They are often used in applications requiring high impact resistance, such as sports equipment and furniture. Bamboo fibers are lightweight and have a high strength-to-weight ratio, making them an excellent choice for reducing the overall weight of composite materials without compromising strength.

2.1.5 Mechanical properties and comparison to synthetic fibers

Natural fibers generally offer a high strength-to-weight ratio, which is comparable to or even superior to some synthetic fibers. For instance, jute fibers have a tensile strength that is comparable to that of glass fibers, but with a significantly lower density. This makes jute fibers an attractive option for reducing the weight of composite materials while maintaining strength [43].

Compared to synthetic fibers, natural fibers offer a more sustainable alternative, with lower environmental impact and enhanced recyclability. Synthetic fibers, such as carbon and glass fibers, are known for their high strength and stiffness but are non-biodegradable and have a higher environmental footprint. Natural fibers, on the other hand, are renewable and biodegradable, and often require less energy to produce.

2.2 Green reinforcements

In composite materials, reinforcements or fillers are added to modify the mechanical, electrical, or thermal characteristics of the base material. These reinforcements can be either organic or inorganic. Traditionally, industries have favored inorganic materials for this purpose, with glass fibers being the most commonly used reinforcement, accounting for approximately 95% of applications [33]. In the rubber industry, nonbiodegradable and potentially harmful fillers like precipitated silica, clay, and calcium carbonate are commonly used [44].

Green reinforcement refers to the use of eco-friendly materials as fillers or reinforcements in composite materials. These materials are primarily derived from natural fibers sourced from plants, such as jute, sisal, coir, hemp, bamboo, kenaf, oil palm, and banana fibers, among others [44]. Natural fibers themselves are considered composite structures, composed of a matrix of hemicellulose and lignin, reinforced by cellulose microfibrils. These microfibrils lend the natural fibers remarkable mechanical properties, including high strength and stiffness. Additionally, natural fibers have a lower density compared to synthetic fibers, and they offer benefits such as biodegradability, renewability, and cost-effectiveness [30, 31]. With growing environmental concerns and the increasing need to reduce the reliance on nonbiodegradable materials, the use of natural fiber reinforcements in composites is rapidly expanding.

Several studies have explored the use of green reinforcements in composite production [44]. Marto and Othman investigated the application of bamboo as a reinforcement for soft clay in embankment construction. They monitored the settlement and lateral movement of three types of embankments: one reinforced with Bamboo-Geotextile Composite (BGC), one with High-Strength Geotextile (HSG), and an unreinforced control (UR). Their results showed that the BGC-reinforced embankment outperformed the others, demonstrating reduced deformation and improved soil stability [45]. Pereira et al. [46] examined the effect of partially replacing Portland cement with rice husk ash (RHA) to enhance the use of green coconut husk fibers as reinforcement in cementitious matrices. Their findings indicated that the substitution of Portland cement with RHA preserved the mechanical properties of the coconut husk fibers, resulting in a lightweight composite with superior mechanical performance [46].

Dominic et al. studied the use of cellulose nanofibers (CNFs) extracted from *Cuscuta reflexa*, a parasitic plant, as fillers in natural rubber (NR). The results showed that incorporating CNFs into NR improved both its processing characteristics and overall performance. This suggests that CNFs could serve as a promising green filler for developing natural rubber-based composites [44]. Zhu et al. utilized waste rubber particles and fly ash from industrial and mining activities to create a new grout material for reinforcing coal seam floors. Their tests revealed that this material exhibited excellent physical and chemical properties, meeting the strength and impermeability requirements for effective floor reinforcement in coal mines [47].

3. Natural fibers

Natural fibers are characterized by structures with a small diameter relative to their length, resulting in a high aspect ratio (length/diameter). While natural fibers can be sourced from plants, minerals, and animals, plant fibers are the most prevalent due to their abundant availability. These fibers are highly valued for their renewability, low density, high strength, elastic modulus, and biodegradability. Plant fibers can be obtained from various plant structures, including stems (e.g., rice, bamboo, corn, wheat, bagasse), leaves (e.g., abaca, pineapple, sisal), seeds (e.g., cotton, kapok, and wider), xylem (e.g., flax and bast), bark (e.g., jute, hibiscus, ramie, and rosella), and fruits (e.g., palm and coconut) [48–50]. The cell walls of these fibers have a complex polysaccharide composition, primarily consisting of cellulose, hemicellulose, and lignin, along with smaller amounts of pectin, waxes, water-soluble substances, and oils. The specific composition of the cell wall varies depending on the plant species, its cell function, developmental stage, and environmental conditions [51]. In terms of structure, plant fibers are typically composed of rigid crystalline cellulose nanofibrils that are aggregated into microfibrils. These microfibrils form slender, rod-like crystalline structures, which are embedded in a matrix of hemicellulose. Lignin acts as a "cement," binding the cellulose microfibrils together and solidifying the cell wall structure [30].

Cellulose is the predominant component in fiber composition and plays a crucial role in providing high tensile strength and Young's modulus, thanks to its crystalline structure [50]. The presence of lignin, however, affects the crystalline structure and typically reduces the fiber's tensile strength. Chemically, cellulose is a hydrophilic glucan polymer, consisting of glucose units connected by β (1 → 4) bonds. It has a high moisture absorption capacity due to the free hydroxyl groups on its surface, which contributes to its hydrophilic nature [52, 53].

Lignin, the second most abundant biopolymer in fiber cell walls, enhances the structural integrity of the fiber. It is chemically bound to hemicellulose through ester bonds and is made up of three primary precursors: coniferyl alcohol (G), p-coumaryl alcohol (H), and sinapyl alcohol (S). The dominant chemical bond in lignin is the aryl ether linkage (β-O-4), which accounts for approximately 50% of the structure [31]. Higher lignin content can typically reduce the mechanical strength of the fiber, but it may enhance resistance to UV degradation and improve the formation of char under high temperatures [31]. Hemicellulose is the third most abundant polysaccharide in the fiber cell wall. It has a branched structure and serves as a link between cellulose and lignin, contributing to the network structure of the fiber. The higher the hemicellulose content, the more it tends to interfere with the crystallinity of cellulose, thus weakening the mechanical properties of the fiber [54]. Additionally, Azwa et al. noted that hemicellulose can improve moisture sorption, which, in turn, affects the biodegradation rate of the fiber [55].

3.1 Cellulose

Cellulose is a macromolecule consisting of nonbranched chains of varying lengths, composed of β-D-anhydroglucopyranose units linked by 1–4 glycosidic bonds. Cellulose pulp, on the other hand, refers to purified cellulosic material that still contains other carbohydrates. The distinction between these terms is significant, as Azwa et al. [56] coined the term "cellulose" to refer to sufficiently purified plant tissues, which today are more accurately called pulp. Other researchers continued to use the original definition by Payen [57], who identified cellulose as a substance similar to starch, composed of glucose residues.

Payen's early experiments, conducted in the mid-nineteenth century, indicated that cellulose was a uniform carbohydrate extracted from fibrous plant tissues like wood, cotton linters, and certain mosses. He hypothesized that cellulose was a highly aggregated form of starch, with a more crystalline structure. This theory, however, faced opposition. For example, Fremy [58], who focused on enzymatic conversions, argued that the properties of cellulose and starch differed because of isomeric variations between the two substances [56]. Payen's analysis of cellulose was based on incomplete experimental data, which led to the misidentification of the pure carbohydrate as being solely composed of glucose. Later, researchers found that cellulose hydrolyzates often contained substantial amounts of other sugars, such as galactose, arabinose, mannose, or xylose, in addition to glucose. In particular, the endosperm of the ivory nut, previously considered to contain pure cellulose, was found to consist almost entirely of mannose [57].

As the understanding of cellulose evolved, the less-resistant carbohydrates other than glucose were grouped under the term "hemicelluloses." Cotton, which contains almost entirely glucose, became the standard reference for cellulose due to its high purity. The term "cellulose," as defined by Payen and later by Zugenmaier [56], now specifically refers to the portion of the plant cell wall resembling cotton cellulose in both physical and chemical properties. Cellulose can also be sourced from various other organisms, including algae (e.g., *Valonia*, *Oocystis apiculata*), tunicates, and even bacteria (*Gluconacetobacter xylinus*). The structure of cellulose can vary significantly depending on its origin.

Cellulose, being responsible for the superior mechanical properties of natural fibers, has become a key focus of research for new applications. Various methods have been developed for extracting cellulose, with chemical techniques being particularly prominent. These include the use of sulfuric acid, followed by chlorination, alkaline treatment, and bleaching [59]. Another approach involves using sodium bisulfate solution in combination with the alkaline treatment [55]. Additionally, acid-chlorite treatment with sodium chlorite acidified with glacial acetic acid is also employed to isolate cellulose [60]. These methods aim to remove the amorphous components in the fiber structure, such as lignin, hemicellulose, and waxes. The resulting product from these treatments is microcellulose (MCC), which consists of microfibrils with both amorphous and crystalline regions held together by van der Waals interactions and hydrogen bonds [61]. The degree of crystallinity of MCC typically ranges from 55 to 80%. MCC is widely used across industries such as pharmaceuticals, food, beverages, and cosmetics, largely due to its hygroscopic properties, which are attributed to the free hydroxyl groups on its surface.

In addition to these applications, composite materials using MCC have been explored due to the inherent mechanical properties of cellulose. However, challenges such as moisture absorption, poor wettability, incompatibility with polymer matrices, and thermal sensitivity [59, 60] still need to be addressed. To overcome these limitations, the development of cellulose nanostructures (CNS) has been investigated. These nanomaterials offer high surface area and aspect ratios, which enhance the intrinsic properties of cellulose. Even at low concentrations, CNS can significantly improve the performance of composite materials, making them suitable for advanced applications in a variety of fields. To overcome these drawbacks, the development of cellulose nanostructures (CNS) has been investigated, since it aligns the material's intrinsic properties with the high surface area and aspect ratio of nanomaterials. Moreover, even at low content, these nanomaterials can considerably increase the properties of composite materials, thus becoming suited for new advanced applications. The isolation of cellulose nanostructures (CNS) can be achieved through acid hydrolysis, enzymatic hydrolysis, or mechanical treatments. The specific morphology, crystallinity, and aspect ratio of the resulting CNS depend on the isolation method employed, as illustrated in **Figure 1**. Among these methods, acid hydrolysis is the most commonly used, typically involving strong acids such as concentrated sulfuric acid. In this process, the acid selectively breaks down the amorphous regions of cellulose, leading to the formation of a stable colloidal dispersion of cellulose nanocrystals

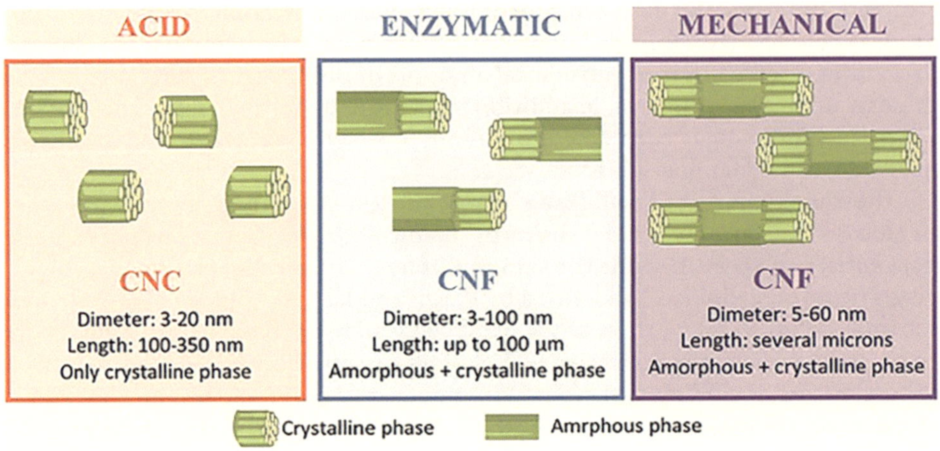

Figure 1.
Different types of cellulose nanostructures and their different properties.

(CNC) or nanowhiskers. These nanocrystals are composed of the crystalline regions that remain intact and typically have a low aspect ratio. While acid hydrolysis is a well-established technique, it generates a large amount of acidic waste, raising significant environmental concerns. As a result, researchers have been exploring alternative methods to reduce the environmental impact associated with this process.

Enzymatic hydrolysis involves the use of a combination of enzymes to promote the scission of cellulose chains. This method operates under mild conditions, making it advantageous in terms of minimizing harsh chemical exposure. However, it requires a longer reaction time compared to other methods, and its cost remains a barrier for large-scale applications. Mechanical treatments, including ultrasonication, ball milling, and high-pressure homogenization, are used to isolate cellulose nanostructures. These processes typically lead to fiber defibrillation, resulting in cellulose nanofibers with an elongated shape and high aspect ratio. The nanofibers produced by mechanical methods exhibit a higher degree of crystallinity than microcrystalline cellulose (MCC), though their crystallinity is still lower than that of cellulose nanocrystals (CNC) obtained through acid hydrolysis. This is because some amorphous regions remain intact after mechanical treatment. Despite these limitations, mechanical treatments offer a promising alternative for producing cellulose nanostructures with desirable properties.

3.1.1 Structure of cellulose

Cellulose consists of polymer chains made up of unbranched, β-(1 → 4)-linked D-glucopyranosyl units, known as anhydroglucose units (AGUs). The length of these glucan chains varies depending on the source of cellulose. In lignocellulosic materials, for instance, the degree of polymerization (DP) can reach up to 10,000. However, these long chains are difficult to quantify directly, as they tend to degrade enzymatically or mechanically during analysis [62, 63]. Each AGU has three hydroxyl groups located at positions C2 and C3 (secondary hydroxyls) and C6 (primary hydroxyl), which can form both intra- and intermolecular hydrogen bonds. These interactions facilitate the formation of highly ordered, three-dimensional crystalline structures.

In vascular plants, the synthesis of glucan chains takes place within specialized transmembrane protein complexes called cellulose synthase complexes (CSCs) or terminal complexes (TCs). A TC is composed of a central globule surrounded by six hexagonally arranged enzyme particles, which together form a rosette approximately 25 nm in diameter. This rosette structure has been confirmed in freeze-fractured plant cell wall samples, observed on the plasmatic face of the plasma membrane. It is now believed that each of the six lobes in the rosette contains six cellulose synthase enzymes, each of which polymerizes a single glucan chain using UDP-glucose as a substrate. These individual chains then align and crystallize to form a cellulose microfibril (MF) with a diameter of about 3.5 nm in wood, typically composed of 36 glucan chains [64]. However, the exact number of active catalytic subunits in the rosette, which determines how many glucan chains are present in a single microfibril, has yet to be conclusively determined and remains a subject of ongoing research [65].

It is important to note that the term "microfibril" refers to a historical concept, originally denoting the smallest isolated entity from the plant cell wall. However, these microfibrils actually exhibit a nanoscale size, typically ranging from 3 to 30 nm, depending on the cellulose's source. The diameter, length, and aspect ratio of various nanofibers derived from cellulose have been documented in several studies (see **Table 1**). A second type of terminal complex has been observed in certain algae and the bacterium *Gluconacetobacter xylinus*, where enzyme arrays are organized in linear patterns, and the number of enzymes varies depending on the organism [66].

The biosynthesis of glucan chains is intricately connected to their assembly and crystallization into highly ordered crystalline domains within the microfibrils. Native cellulose (cellulose I) exists in two distinct crystalline forms, referred to as cellulose Iα and Iβ, with the relative proportions of these forms varying based on the cellulose's origin. Cellulose Iα consists of triclinic unit cells, whereas cellulose Iβ, the dominant form in higher plants, exhibits a monoclinic unit cell structure.

Cellulose II, an allomorph of cellulose, is rarely found in nature, with examples such as the marine algae Halicystis. However, it can be synthesized from cellulose I through two processes: regeneration and mercerization. The regeneration process involves dissolving cellulose in a specific solvent, like N-methylmorpholine-N-oxide, while mercerization involves swelling cellulose in an aqueous sodium hydroxide solution. In both cases, a final re-crystallization step transforms cellulose I into cellulose II, which is thermodynamically more stable than cellulose I. Notably, evidence suggests that cellulose II consists of antiparallel glucan chains, in contrast to the parallel arrangement found in cellulose I [65], though this remains an area of active debate. In addition to celluloses I and II, other cellulose allomorphs have been identified, including cellulose III and cellulose IV [62, 63]. The glucan chains of cellulose synthases assemble into a highly ordered structure, forming microfibrils.

Cellulose structure	Diameter (nm)	Length (nm)	Aspect ratio (L/d)
Microfibril	2–10	>10,000	>1000
Microfibrillated cellulose (MFC)	10–40	>1000	100–150
Cellulose whisker	2–20	100–600	10–100
Microcrystalline cellulose (MCC)	>1000	>1000	~1

Table 1.
Nanocellulose dimensions.

These microfibrils, however, are not entirely crystalline; they contain para-crystalline (amorphous) regions with lower order and defects. The widely accepted model to describe this structure is the fringed-fibrillar model, which proposes that the glucan chains move irregularly between amorphous and crystalline domains [62]. During acidic hydrolysis, the glucan chains are more likely to break in the amorphous regions, producing microfibril fragments known as whiskers due to their slender, rod-like appearance. These whiskers can then aggregate into larger fibril bundles or agglomerates, held together by matrix substances like hemicelluloses, lignin, and pectin.

As the structural backbone of plants, cellulose is organized within a hierarchical cellular structure. Plant cell walls are separated by a middle lamella and are made up of a primary cell wall layer. The secondary cell wall is further divided into layers S1 and S2, with the latter containing the majority of the cellulose [62]. The cellulose microfibrils within these cell walls are oriented at characteristic angles, which vary depending on the plant type and the specific layer of the cell wall. This orientation is likely influenced by microtubules, which often align parallel to the microfibrils. During the biosynthesis of glucan chains, it is believed that the terminal complexes (TCs) move backward, driven by the rigidity of the microfibrils, and this motion is restricted by the channels formed by the oriented microtubules [66]. The orientation of cellulose microfibrils significantly influences the mechanical properties of plant fibers. For example, in the S2 layer, where the microfibrils are oriented nearly parallel to the fiber axis, the fiber exhibits a high modulus of elasticity. In contrast, larger microfibril angles result in greater elongation at break, indicating more flexibility [62]. This structural arrangement, coupled with the abundance of hydrogen bonds within the fibrillar structure, gives cellulose its exceptional tensile strength. As a result, cellulose plays a crucial role as the primary structural component in plants, particularly in bearing tensile loads [67].

3.1.2 Surface modification

While cellulose nanostructures (CNS) can be directly used to create nanocomposites, they often face compatibility issues at the polymeric interface. This is primarily due to the polarity differences between the filler (CNS) and the polymer matrix. Each glucose unit in the cellulose chain has three hydroxyl groups that can form strong hydrogen bonds between chains, promoting agglomeration of the cellulose nanostructures. Additionally, the hydrophobic nature of many polymer chains reduces their interaction with CNS, limiting the potential for enhancing the composite's properties. To overcome these challenges, several surface modification techniques can be employed. These include silylation, esterification, etherification, oxidation, carbamation, and polymer grafting. These treatments target the hydroxyl groups on the surface of CNS, introducing various chemical structures that improve their dispersion in the matrix and enhance their interaction with the polymer. Such modifications significantly improve the compatibility between CNS and the polymeric matrix, allowing for more efficient reinforcement and better overall performance of the resulting nanocomposite. Esterification, such as acetylation, is commonly used to modify cellulose nanostructures (CNS). This process introduces ester groups onto the surface of CNS, which reduces their hydrophobic nature and improves compatibility with hydrophilic matrices. Acetylation, specifically, utilizes acetic acid and acetic anhydride to attach acetyl groups to the CNS surface [68]. The effectiveness of this modification is measured by the degree of substitution (DS), which indicates the average number of hydroxyl groups that have been replaced by acetyl groups on the cellulose structure.

Silylation, another modification technique, involves the incorporation of alkoxy silane groups onto the surface of CNS. This enhances the wettability of the fibers with the polymer matrix, improving the dispersion and interaction between the filler and the matrix [69]. Carbamation introduces isocyanate groups, such as 2,4-diisocyanate toluene (TDI), hexamethylene diisocyanate (HMDI), and methylene diphenyl diisocyanate, onto the CNS surface. These isocyanate groups can serve as chemical linkers in subsequent reactions, facilitating the formation of stronger bonds between CNS and the polymeric matrix [70]. Polymer grafting is another approach that increases the stability of CNS and improves their interaction with the polymeric matrix. This method involves introducing polymer chains onto the cellulose surface, either through direct polymerization or by using a coupling agent. This modification allows for the attachment of various structures, enhancing the compatibility and performance of the resulting composite material [71].

4. Composite processing

As natural fibers at the nanoscale demonstrate attractive properties, there is growing interest in incorporating these nanofillers into plastic matrices as functional materials. This approach helps reduce the consumption of synthetic polymer materials, encourages the use of renewable raw materials, and lowers production costs and environmental impacts. It also facilitates the development of products with improved biodegradability, as shown in **Figure 2**. Polymer composites containing natural fibers can be produced using conventional polymer processing techniques, making them suitable for large-scale production. The methods commonly used for preparing fiber-based composites include melt blending (such as extrusion and injection molding), coating, roll-to-roll coating and casting, and solvent casting [72].

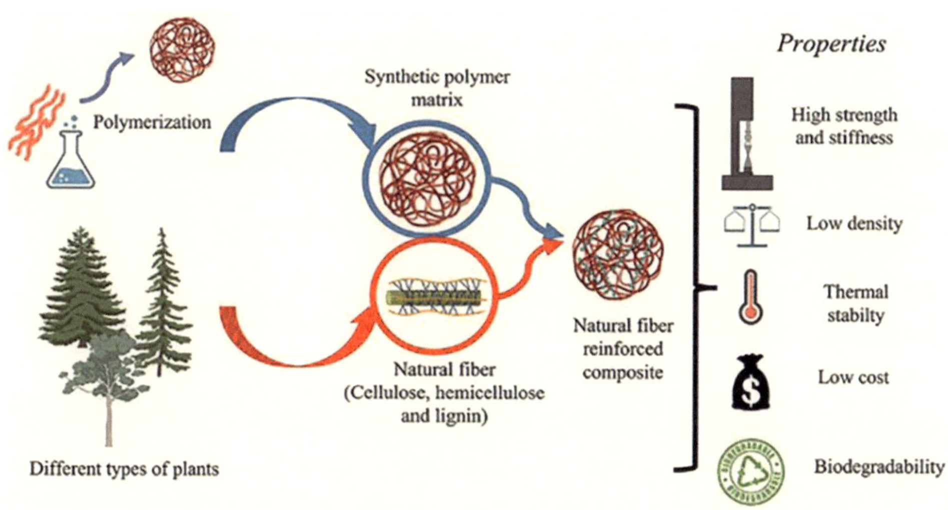

Figure 2.
Structural properties found in natural fiber composites with biopolymers.

5. Conclusion

Biobased composites represent a significant advancement in the quest for sustainable materials, offering a viable alternative to traditional petroleum-based products. By integrating natural fibers with biopolymers, these composites not only reduce environmental impact but also enhance material properties such as strength, stiffness, and thermal stability. Their applications across various industries, including automotive, construction, and consumer goods, underscore their versatility and potential to drive a shift toward more sustainable practices. As research continues to optimize processing techniques and improve material properties, biobased composites are poised to play a crucial role in the circular economy, promoting resource efficiency and waste reduction. However, the long-term durability of biobased composites remains an area that requires further investigation. While current advancements are promising, a deeper understanding of their performance over time is essential to ensure their reliability and effectiveness in diverse applications. Addressing these knowledge gaps will be critical in fully realizing the potential of biobased composites and in supporting global efforts to reduce carbon footprints and promote environmental stewardship. As consumer demand for eco-friendly products grows, the development of robust, high-performance biobased composites will be key to achieving a more sustainable future.

The advancements in biocomposites underscore their potential to replace traditional petroleum-based materials across various industries. With improved mechanical properties, such as enhanced strength and stiffness, alongside their biodegradability, biocomposites present a compelling solution to the environmental challenges posed by conventional materials. Continued research and development in this field will be crucial for optimizing their performance and expanding their applications.

Acknowledgements

The authors appreciate the support from the University of Mazandaran. Also, the generous contributions of all our collaborators over the years are appreciated.

Conflict of interest

The authors declare no conflict of interest.

DOI: http://dx.doi.org/10.5772/intechopen.1009464

Author details

Behrooz Maleki*, Pouya Ghamari Kargar and Samaneh Sedigh Ashrafi
Faculty of Chemistry, Department of Organic Chemistry, University of Mazandaran, Babolsar, Iran

*Address all correspondence to: malekibehrooz@gmail.com

References

[1] Wu F, Misra M, Mohanty AK. Challenges and new opportunities on barrier performance of biodegradable polymers for sustainable packaging. Progress in Polymer Science [Internet]. 2021;**117**:101395. DOI: 10.1016/j.progpolymsci.2021.101395

[2] Amulya K, Katakojwala R, Ramakrishna S, Venkata MS. Low carbon biodegradable polymer matrices for sustainable future. Composites Part C: Open Access [Internet]. 2021;**4**:100111. DOI: 10.1016/j.jcomc.2021.100111

[3] Musto P. Grand challenges in polymer chemistry: Energy, environment, health. Frontiers in Chemistry [Internet]. 2013;**1**:31. DOI: 10.3389/fchem.2013.00031

[4] Hameed M, Bhat RA, Singh DV, Mehmood MA. White Pollution: A Hazard to Environment and Sustainable Approach to its Management. IGI Global Scitific Publishing [Internet]; 2020. pp. 52-81. DOI: 10.4018/978-1-7998-0031-6.ch004

[5] George A, Sanjay MR, Srisuk R, Parameswaranpillai J, Siengchin S. A comprehensive review on chemical properties and applications of biopolymers and their composites. International Journal of Biological Macromolecules [Internet]. 2020;**154**:329-338. DOI: 10.1016/j.ijbiomac.2020.03.120

[6] Khan MZ, Srivastava SK, Gupta M. Tensile and flexural properties of natural fiber reinforced polymer composites: A review. Journal of Reinforced Plastics and Composites [Internet]. 2018;**37**(24):1435-1455. DOI: 10.1177/0731684418799528

[7] Ferreira RR, Souza AG, Rosa DS. Essential oil-loaded nanocapsules and their application on PBAT biodegradable films. Journal of Molecular Liquids [Internet]. 2021;**337**:116488. DOI: 10.1016/j.molliq.2021.116488

[8] Vikhareva IN, Buylova EA, Yarmuhametova GU, Aminova GK, Mazitova AK. An overview of the main trends in the creation of biodegradable polymer materials. Journal of Chemistry [Internet]. 2021;**2021**:1-15. DOI: 10.1155/2021/5099705

[9] Chen C, Xi Y, Weng Y. Recent advances in cellulose-based hydrogels for tissue engineering applications. Polymers (Basel) [Internet]. 2022;**14**(16):3335. DOI: 10.3390/polym14163335

[10] Mohanapriya MK, Deshmukh K, Chidambaram K, Ahamed MB, Sadasivuni KK, Ponnamma D, et al. Polyvinyl alcohol (PVA)/polystyrene sulfonic acid (PSSA)/carbon black nanocomposite for flexible energy storage device applications. Journal of Materials Science: Materials in Electronics [Internet]. 2017;**28**(8):6099-6111. DOI: 10.1007/s10854-016-6287-2

[11] Deshmukh K, Ahamed MB, Sadasivuni KK, Ponnamma D, Deshmukh RR, Trimukhe AM, et al. Solution-processed white graphene-reinforced ferroelectric polymer nanocomposites with improved thermal conductivity and dielectric properties for electronic encapsulation. Journal of Polymer Research [Internet]. 2017;**24**(2):27. DOI: 10.1007/s10965-017-1189-4

[12] Ferreira F, Pinheiro I, de Souza S, Mei L, Lona L. Polymer composites reinforced with natural fibers and nanocellulose in the automotive industry: A short review. Journal of Composites

Science [Internet]. 2019;**3**(2):51. DOI: 10.3390/jcs3020051

[13] Bafakeeh OT, Shewakh WM, Abu-Oqail A, Abd-Elaziem W, Abdel Ghafaar M, Abu-Okail M. Synthesis and characterization of hybrid fiber-reinforced polymer by adding ceramic nanoparticles for aeronautical structural applications. Polymers (Basel) [Internet]. 2021;**13**(23):4116. DOI: 10.3390/polym13234116

[14] Kirillova A, Yeazel TR, Asheghali D, Petersen SR, Dort S, Gall K, et al. Fabrication of biomedical scaffolds using biodegradable polymers. Chemical Reviews [Internet]. 2021;**121**(18):11238-11304. DOI: 10.1021/acs.chemrev.0c01200

[15] Ferreira FV, Dufresne A, Pinheiro IF, Souza DHS, Gouveia RF, Mei LHI, et al. How do cellulose nanocrystals affect the overall properties of biodegradable polymer nanocomposites: A comprehensive review. European Polymer Journal [Internet]. 2018;**108**:274-285. DOI: 10.1016/j.eurpolymj.2018.08.045

[16] Abdillah AA, Charles AL. Characterization of a natural biodegradable edible film obtained from arrowroot starch and iota-carrageenan and application in food packaging. International Journal of Biological Macromolecules [Internet]. 2021;**191**:618-626. DOI: 10.1016/j.ijbiomac.2021.09.141

[17] Khoo RZ, Chow WS, Ismail H. Tensile, thermal and ultra-violet shielding enhancement of poly(lactic acid) bionanocomposite film using cellulose nanocrystals extracted from sugarcane bagasse. Journal of Thermoplastic Composite Materials [Internet]. 2023;**36**(6):2543-2561. DOI: 10.1177/08927057221098972

[18] Udayakumar GP, Muthusamy S, Selvaganesh B, Sivarajasekar N, Rambabu K, Banat F, et al. Biopolymers and composites: Properties, characterization and their applications in food, medical and pharmaceutical industries. Journal of Environmental Chemical Engineering [Internet]. 2021;**9**(4):105322. DOI: 10.1016/j.jece.2021.105322

[19] Priyadarshi R, Roy S, Ghosh T, Biswas D, Rhim JW. Antimicrobial nanofillers reinforced biopolymer composite films for active food packaging applications - a review. Sustainable Materials and Technologies [Internet]. 2022;**32**:e00353. DOI: 10.1016/j.susmat.2021.e00353

[20] Rajeswari S. Natural polymers: A recent review. World Journal Pharmacy and Pharmaceutical Science [Internet]. 2017;**8**:472-494. Available from: http://wjpps.com/wjpps_controller/abstract_id/7480

[21] Sarikaya E, Çallioğlu H, Demirel H. Production of epoxy composites reinforced by different natural fibers and their mechanical properties. Composites Part B: Engineering [Internet]. 2019;**167**:461-466. DOI: 10.1016/j.compositesb.2019.03.020

[22] Sivakanthan S, Rajendran S, Gamage A, Madhujith T, Mani S. Antioxidant and antimicrobial applications of biopolymers: A review. Food Research International [Internet]. 2020;**136**:109327. DOI: 10.1016/j.foodres.2020.109327

[23] Mitura S, Sionkowska A, Jaiswal A. Biopolymers for hydrogels in cosmetics: Review. Journal of Materials Science: Materials in Medicine [Internet]. 2020;**31**(6):50. DOI: 10.1007/s10856-020-06390-w

[24] Li X, Ding C, Li X, Yang H, Liu S, Wang X, et al. Electronic biopolymers: From molecular engineering to functional devices. Chemical Engineering Journal [Internet]. 2020;**397**:125499. DOI: 10.1016/j.cej.2020.125499

[25] Hayes DG, Dharmalingam S, Wadsworth LC, Leonas KK, Miles C, Inglis DA. Biodegradable agricultural mulches derived from biopolymers. In: Degradable Polymers and Materials: Principles and Practice. 2nd ed. [Internet]. 2012;**1114**:201-223. DOI: 10.1021/bk-2012-1114.ch013

[26] Udayakumar GP, Muthusamy S, Selvaganesh B, Sivarajasekar N, Rambabu K, Sivamani S, et al. Ecofriendly biopolymers and composites: Preparation and their applications in water-treatment. Biotechnology Advances [Internet]. 2021;**52**:107815. DOI: 10.1016/j.biotechadv.2021.107815

[27] Sawant SN. Development of biosensors from biopolymer composites. In: Biopolymer Composites in Electronics [Internet]. Elsevier; 2017. pp. 353-383. DOI: 10.1016/B978-0-12-809261-3.00013-9

[28] Avérous L, Pollet E. Environmental Silicate Nano-Biocomposites [Internet]. London: Springer London; 2012;**1**:1-450. DOI: 10.1007/978-1-4471-4108-2

[29] Kulkarni Vishakha S, Butte Kishor D, Rathod SS. Natural polymers–a comprehensive review. International Journal of Research in Pharmaceutical and Biomedical Sciences. 2012;**4**(3):1597-1613

[30] Djafari, Petroudy SR. Physical and mechanical properties of natural fibers. In: Advanced High Strength Natural Fibre Composites in Construction. Woodhead Publishing [Internet]; 2017:59-83. DOI: 10.1016/B978-0-08-100411-1.00003-0

[31] Karimah A, Ridho MR, Munawar SS, Adi DS, Ismadi DR, et al. A review on natural fibers for development of eco-friendly bio-composite: Characteristics, and utilizations. Journal of Materials Research and Technology [Internet]. 2021;**13**:2442-2458. DOI: 10.1016/j.jmrt.2021.06.014

[32] Asim M, Paridah MT, Chandrasekar M, Shahroze RM, Jawaid M, Nasir M, et al. Thermal stability of natural fibers and their polymer composites. Iranian Polymer Journal [Internet]. 2020;**29**(7):625-648. DOI: 10.1007/s13726-020-00824-6

[33] Hsissou R, Seghiri R, Benzekri Z, Hilali M, Rafik M, Elharfi A. Polymer composite materials: A comprehensive review. Composite Structures [Internet]. 2021;**262**:113640. DOI: 10.1016/j.compstruct.2021.113640

[34] Institution of Engineering and Technology. Engineering competency. In: High Voltage Power Network Construction [Internet]. 2024:689-703. DOI: 10.1049/pbpo110e_ch26

[35] Mohammed L, Ansari MNM, Pua G, Jawaid M, Islam MS. A review on natural fiber reinforced polymer composite and its applications. International Journal of Polymer Science [Internet]. 2015;**2015**: 1-15. DOI: 10.1155/2015/243947

[36] Roy K, Debnath SC, Pongwisuthiruchte A, Potiyaraj P. Recent advances of natural fibers based green rubber composites: Properties, current status, and future perspectives. Journal of Applied Polymer Science [Internet]. 2021;**138**(35):50866. DOI: 10.1002/app.50866

[37] Adeniyi AG, Onifade DV, Ighalo JO, Adeoye AS. A review of coir fiber

reinforced polymer composites. Composites Part B: Engineering [Internet]. 2019;**176**:107305. DOI: 10.1016/j.compositesb.2019.107305

[38] Sanjay MR, Madhu P, Jawaid M, Senthamaraikannan P, Senthil S, Pradeep S. Characterization and properties of natural fiber polymer composites: A comprehensive review. Journal of Cleaner Production [Internet]. 2018;**172**:566-581. DOI: 10.1016/j.jclepro.2017.10.101

[39] Ramasubbu R, Madasamy S. Fabrication of automobile component using hybrid natural fiber reinforced polymer composite. Journal of Natural Fibers [Internet]. 2022;**19**(2):736-746. DOI: 10.1080/15440478.2020.1761927

[40] Otto GP, Moisés MP, Carvalho G, Rinaldi AW, Garcia JC, Radovanovic E, et al. Mechanical properties of a polyurethane hybrid composite with natural lignocellulosic fibers. Composites Part B: Engineering [Internet]. 2017;**110**:459-465. DOI: 10.1016/j.compositesb.2016.11.035

[41] Ashik KP, Sharma RS, Guptha VLJ. Investigation of moisture absorption and mechanical properties of natural/glass fiber reinforced polymer hybrid composites. Materials Today: Proceedings [Internet]. 2018;**5**(1):3000-3007. DOI: 10.1016/j.matpr.2018.01.099

[42] Omrani E, Menezes PL, Rohatgi PK. State of the art on tribological behavior of polymer matrix composites reinforced with natural fibers in the green materials world. Engineering Science and Technology, an International Journal. [Internet]. 2016;**19**(2):717-736. DOI: 10.1016/j.jestch.2015.10.007

[43] Elfaleh I, Abbassi F, Habibi M, Ahmad F, Guedri M, Nasri M, et al. A comprehensive review of natural fibers and their composites: An eco-friendly alternative to conventional materials. Results in Engineering [Internet]. 2023;**19**:101271. DOI: 10.1016/j.rineng.2023.101271

[44] Dominic CDM, Joseph R, Begum PMS, Joseph M, Padmanabhan D, Morris LA, et al. Cellulose nanofibers isolated from the *Cuscuta reflexa* plant as a green reinforcement of natural rubber. Polymers (Basel) [Internet]. 2020;**12**(4):814. DOI: 10.3390/polym12040814

[45] Marto A, Othman BA. The potential use of bamboo as green material for soft clay reinforcement system. International Conference on Environment Science and Engineering Singapore. 2011;**8**:129-133

[46] Pereira CL, Savastano H, Payá J, Santos SF, Borrachero MV, Monzó J, et al. Use of highly reactive rice husk ash in the production of cement matrix reinforced with green coconut fiber. Industrial Crops and Products [Internet]. 2013;**49**:88-96. DOI: 10.1016/j.indcrop.2013.04.038

[47] Zhu X, Zhang Q, Zhang W, Shao J, Wang Z, Wu X. Experimental study on the basic properties of a green new coal mine grouting reinforcement material. ACS Omega [Internet]. 2020;**5**(27):16722-16732. DOI: 10.1021/acsomega.0c01626

[48] Corrente GA, Scarpelli F, Caputo P, Rossi CO, Crispini A, Chidichimo G, et al. Chemical–physical and dynamical–mechanical characterization on *Spartium junceum* L. cellulosic fiber treated with softener agents: A preliminary investigation. Scientific Reports [Internet]. 2021;**11**(1):35. DOI: 10.1038/s41598-020-79568-5

[49] Sari NH, Fajrin J, Suteja FA. Characterisation of swellability and

compressive and impact strength properties of corn husk fibre composites. Composites Communications [Internet]. 2020;**18**:49-54. DOI: 10.1016/j.coco.2020.01.009

[50] Joshi S, Drzal L, Mohanty A, Arora S. Are natural fiber composites environmentally superior to glass fiber reinforced composites? Composites Part A: Applied Science and Manufacturing [Internet]. 2004;**35**(3):371-376. DOI: 10.1016/j.compositesa.2003.09.016

[51] Petrova A, Kozlova L, Gorshkov O, Nazipova A, Ageeva M, Gorshkova T. Cell wall layer induced in xylem fibers of flax upon gravistimulation is similar to constitutively formed cell walls of bast fibers. Frontiers in Plant Science [Internet]. 2021;**12**:660375. DOI: 10.3389/fpls.2021.660375/full

[52] Zhai L, Kim HC, Kim JW, Kang J, Kim J. Elastic moduli of cellulose nanofibers isolated from various cellulose resources by using aqueous counter collision. Cellulose [Internet]. 2018;**25**(7):4261-4268. DOI: 10.1007/s10570-018-1836-x

[53] Omran AAB, Mohammed AABA, Sapuan SM, Ilyas RA, Asyraf MRM, Rahimian Koloor SS, et al. Micro- and nanocellulose in polymer composite materials: A review. Polymers (Basel) [Internet]. 2021;**13**(2):231. DOI: 10.3390/polym13020231

[54] Ahankari SS, Subhedar AR, Bhadauria SS, Dufresne A. Nanocellulose in food packaging: A review. Carbohydrate Polymers [Internet]. 2021;**255**:117479. DOI: 10.1016/j.carbpol.2020.117479

[55] Wan J, Wang Y, Xiao Q. Effects of hemicellulose removal on cellulose fiber structure and recycling characteristics of eucalyptus pulp. Bioresource Technology [Internet]. 2010;**101**(12):4577-4583. DOI: 10.1016/j.biortech.2010.01.026

[56] Azwa ZN, Yousif BF, Manalo AC, Karunasena W. A review on the degradability of polymeric composites based on natural fibres. Materials & Design [Internet]. 2013;**47**:424-442. DOI: 10.1016/j.matdes.2012.11.025

[57] Zugenmaier P. Crystalline Cellulose and Derivatives [Internet], Springer Series in Wood Science. Berlin, Heidelberg: Springer Berlin Heidelberg; 2008. DOI: 10.1007/978-3-540-73934-0

[58] Zugenmaier P. Purves. Introduction. In: Crystalline Cellulose and Derivatives. Berlin, Heidelberg: Springer; 2008. pp. 1-6. DOI: 10.1007/978-3-540-73934-0_1

[59] Fremy E. Remarques a l'occasion d'une communication de M Payen sur les tissus des. vegetaux. Comptes Rendus. Académie des Sciences. 1859;**48**:325-360

[60] Bolio-López GI, Cadenas-Madrigal G. Extraction of cellulose fibers from tó leaf petioles (*Calathea lutea*) and characterization. International Journal of Innovative Science Engineering and Technology. 2015;**2**:977-981

[61] Galiwango E, Abdel Rahman NS, Al-Marzouqi AH, Abu-Omar MM, Khaleel AA. Isolation and characterization of cellulose and α-cellulose from date palm biomass waste. Heliyon [Internet]. 2019;**5**(12):e02937. DOI: 10.1016/j.heliyon.2019.e02937

[62] Trache D, Hussin MH, Hui Chuin CT, Sabar S, Fazita MRN, Taiwo OFA, et al. Microcrystalline cellulose: Isolation, characterization and bio-composites application—A review. International Journal of Biological Macromolecules [Internet]. 2016;**93**:789-804. DOI: 10.1016/j.ijbiomac.2016.09.056

[63] Klemm D, Heublein B, Fink H, Bohn A. Cellulose: Fascinating biopolymer and sustainable raw material. Angewandte Chemie International Edition [Internet]. 2005;**44**(22):3358-3393. DOI: 10.1002/anie.200460587

[64] O'sullivan AC. Cellulose: The structure slowly unravels. Cellulose. 1997;**3**(4):173-207. DOI: 10.1023/A:1018431705579

[65] Diotallevi F, Mulder B. The cellulose synthase complex: A polymerization driven supramolecular motor. Biophysical Journal [Internet]. 2007;**92**(8):2666-2673. DOI: 10.1529/biophysj.106.099473

[66] Bessueille L, Bulone V. A survey of cellulose biosynthesis in higher plants. Plant Biotechnology [Internet]. 2008;**25**(3):315-322. DOI: 10.5511/plantbiotechnology.25.315

[67] Delmer DP. Cellulose biosynthesis. Annual Review of Plant Physiology and Plant Molecular Biology [Internet]. 1987;**38**(1):259-290. DOI: 10.1146/annurev.pp.38.060187.001355

[68] Sjöström E. Wood Chemistry: Fundamentals and Applications. 2nd ed. San Diego: CA Acad Press Inc. Elsevier; 1993

[69] Abdelmouleh M, Boufi S, Belgacem MN, Dufresne A, Gandini A. Modification of cellulose fibers with functionalized silanes: Effect of the fiber treatment on the mechanical performances of cellulose–thermoset composites. Journal of Applied Polymer Science [Internet]. 2005;**98**(3):974-984. DOI: 10.1002/app.22133

[70] Zheng X, Song Z, Liu E, Zhang Y, Li Z. Preparation of phosphoric acid-functionalized SBA-15 and its high efficient selective adsorption separation of lanthanum ions. Journal of Chemical & Engineering Data [Internet]. 2020;**65**(2):746-756. DOI: 10.1021/acs.jced.9b00976

[71] Sharma A, Thakur M, Bhattacharya M, Mandal T, Goswami S. Commercial application of cellulose nano-composites – A review. Biotechnology Reports [Internet]. 2019;**21**:e00316. DOI: 10.1016/j.btre.2019.e00316

[72] Das TK, Ghosh P, Das NC. Preparation, development, outcomes, and application versatility of carbon fiber-based polymer composites: A review. Advanced Composites and Hybrid Materials [Internet]. 2019;**2**(2):214-233. DOI: 10.1007/s42114-018-0072-z